CLIMATE POLITICS AND
THE POWER OF RELIGION

CLIMATE POLITICS AND THE POWER OF RELIGION

Edited by Evan Berry

INDIANA UNIVERSITY PRESS

This book is a publication of

Indiana University Press
Office of Scholarly Publishing
Herman B Wells Library 350
1320 East 10th Street
Bloomington, Indiana 47405 USA

iupress.org

Manufactured in the United States of America

First printing 2022

Cataloging information is available from the Library of Congress.

ISBN 978-0-253-05905-5 (hardback)
ISBN 978-0-253-05906-2 (paperback)
ISBN 978-0-253-05907-9 (ebook)

CONTENTS

PREFACE

FOR FIFTY YEARS, SCHOLARS FROM THE HUMANITIES AND social sciences have sought to better understand the role of religion in ecological issues. Research in this field has tended to cluster around a set of conventions that crystalized early in the formation of this field. First, debates have tended to focus on whether religion generally or particular religious traditions are "good" or "bad" for the environment. Second, the bulk of published literature has tended to concentrate on North Atlantic societies and especially on the forms of Christianity predominant in the United States, Canada, and Western Europe. These foci have remained as constraints on the emerging body of scholarship on religion and climate change.

Climate Politics and the Power of Religion and its companion volume, *Understanding Climate Change through Religious Lifeworlds*, were outcomes of a research project that aimed to broaden the conversation about religion and climate change by expanding the geographic frame of reference, thinking comparatively, and emphasizing ethnographic scholarship. Funded by the Henry Luce Foundation's Religion and International Affairs Program and managed by American University's Center for Latin American and Latino Studies (CLALS), this project was called "Religion and Climate Change in Cross-Regional Perspective." Although they do not appear extensively in the pages that follow, Eric Hershberg, Director of CLALS, and Rob Albro, CLALS Research Associate Professor, played vital roles in shaping the conversations that formed the basis for this book.

Religion and Climate Change in Cross-Regional Perspective began with a focus on three highly visible forms of environmental vulnerability to the impacts of climate change. The project hypothesized that religious communities around the world are confronted by similar environmental pressures and that their adaptive responses might afford comparative insight about the ways in which religion comes to matter with respect to climate change. Specifically, the project was structured around a series of workshops, each of which foregrounded cases from one front-line impact of climate change. A 2016 workshop at the Observer Research Foundation in Delhi, India, concentrated on urban water scarcity and the challenges faced by megacities in South Asia and South America as they struggle with issues of supply and sanitation. The second workshop was held at Universidad Antonio Ruiz de Montoya in Lima, Peru, in 2017 and focused on glacial melt and the ecological precarity of high-elevation communities in the Andes and in the Himalayas. The final workshop, convened at the University of the West Indies, St. Augustine, in Trinidad and Tobago in 2017, was dedicated to the special challenges facing small island developing states in the Caribbean basin and in the South Pacific—namely, sea level rise and storm intensification. These workshops provided space for contributors to share, refine, and recalibrate their scholarship in conversation with experts from other disciplinary and regional contexts.

Exchange among project contributors was further facilitated by a provisional framework regarding the role of religion: conversations were organized into three streams, each representing one important mode of engagement between religion and climate change. In the first, contributors examined the significance of religious actors, including faith-based organizations, religious leaders, and institutionalized systems of religious mobilization. The second explored religious frames of reference, seeking knowledge about the ways in which religious beliefs, perceptions, and vocabularies shape the way human communities articulate and engage the diverse range of phenomena invoked within the perhaps too comprehensive term *climate change*. The final stream imagined the relationship between religion and climate change differently, asking whether, when, and how environmental changes precipitate religious changes.

These complex questions have no easy answers. The roughly two dozen scholarly essays that emerged from the Religion and Climate Change in Cross-Regional Perspective project suggest some of the many ways these intersections can be seen and understood, it is hoped, in ways that evoke sympathy and humane policy responses.

CLIMATE POLITICS AND
THE POWER OF RELIGION

INTRODUCTION

Evan Berry

S INCE THE ESTABLISHMENT OF A CLIMATE-SPECIFIC APPARATUS FOR international diplomatic engagement in the early 1990s, climate change has
moved from the geopolitical periphery to its epicenter. The collective impact of human activity on the global environment now figures prominently
in virtually all domains of public life, from economic development to military strategy, from cultural production to international relations. No longer
merely a speculative concern about future generations, climate change has
infused twenty-first–century politics with the raw facts of environmental
precarity. One might even say that variations in national politics can be
characterized by the degree to which such precarity has been absorbed
into popular consciousness and public policy—neither the resurgence of
populist nationalisms nor the renewed enthusiasm for the "green left" in
Europe and the Americas can be disentangled from the looming climate
crisis. Contestations about trade policy, transportation, migration, wildlife
conservation, and agriculture are everywhere imbued with climate anxieties. Some nations have contributed much more than others to the problem,
and some nations are much more vulnerable than others to the deleterious
impacts of climate change. Scholars and policy makers tend to discuss such
inequalities by using the language of justice. They draw on diplomatic, cosmopolitan vocabularies to articulate why the world is warming and what
should be done to ameliorate this expanding, slow-motion planetary disaster. On this account, climate change is an inherently global phenomenon;
the singular ecological challenge that humanity now faces joins us together
in a planetary polity. This is the banal Anthropocene—an acknowledgment
of the collective impact of human beings on the Earth's ecology without
serious introspection about the human condition and the manifold ways of
being human on our warming planet.

It is true that climate change is a global justice issue. But the cosmopolitan frame within which experts and elites often situate the climate crisis does not fully appreciate the complex dynamics of climate politics at scales below the global. At the national and subnational levels, public discourse and policy debates do not necessarily replicate the logics that hold sway among international elites. Despite its attention to the uneven distribution of harms, cosmopolitan approaches can have a flattening effect, failing to capture the texture and specificity with which environmental change is experienced. Anthropogenic disruption of the atmosphere's chemistry cannot be understood simply in terms of generic harms being dispersed across time and space. Global warming is, ironically, predicated on causes that are not genuinely global and, likewise, produces consequences that are not universal. There is tremendous variation in the ways that societies understand, contest, and respond to the host of large-scale changes collectively referred to as *climate change*.[1] To take seriously the politics of climate change involves not only a commitment to global justice but also an insistence on particularity, autonomy, and difference. It matters how different communities render climate change within their own moral vocabularies, and it matters whether and how such moral claims find purchase in public debates about climate policy. *Climate Politics and the Power of Religion* aims to foster deeper understanding of such variation by attending to the role of religion in shaping climate politics at the local, national, and international levels. Throughout this volume, attention to religion provides a lens through which the politics of environmental precarity appear in more vivid, particularistic detail.

This approach—taken here by a group of scholars employing a range of disciplinary methods and perspectives—aims to counter the secular bias latent in much of the scholarly literature on environmental politics. *Climate Politics and the Power of Religion* is premised on the idea that religion is an important factor in public conversations about climate change, that it plays a key role in shaping national-level discourse about climate policy, and that these religious aspects are refracted into international climate diplomacy and multilateral policy contexts. The term *religion* is, to say the least, complicated. It is used to denote a variety of social phenomena, including the

---

1. The companion to this volume, David Haberman, ed., *Understanding Climate Change and Religious Lifeworlds* (Bloomington: Indiana University Press, 2021), takes this approach, documenting how different societies' perceptions of and responses to climate change are refracted through religious lenses.

kinds of cultural traditions that shape collective ideas about the natural order, scriptural and theological resources for discerning ethical action, distinct types of actors in the public sphere, and forms of discourse that speak gods into the human world. In conjoining these different approaches to the category of religion, this volume aims to provide an interdisciplinary framework for expanding academic and public interest in the intersection of religion and climate change.

Although formulated in distinct disciplinary terminologies, the interface of religion and climate change has attracted attention from political scientists, sociologists, anthropologists, historians, theologians, scholars of religion, geographers, literary critics, and indigenous studies scholars. Nevertheless, despite this growing interest in religion and climate change, few attempts have been made to grapple with this intersection systematically or comparatively (notable exceptions include Gerten and Bergmann 2012; Veldman, Szasz, and Haluza-DeLay 2013; Haluza-Delay 2014; LeVasseur 2015). As is frequently and unfortunately the case, the bulk of what has been published on this topic focuses on the United States and other wealthy nations in the global North. The climate crisis is a spiritual crisis, and the existential anxieties this provokes among the culpable societies of the global North have been well documented (e.g., Norgaard 2011; Willox 2012; Sponsel 2012; Pikhala 2018). Religious responses to climate change in the global South are less well known (see Duara 2015).

Climate Politics and the Power of Religion addresses a set of related knowledge gaps: a lack of empirical information about religion and climate change in poorer societies, especially where Christianity is not the dominant religious tradition; a lack of attention to religion among the core social scientific literatures on climate change; and a lack of reflection about politics in much of the literature on religion and the environment. This volume is the result of a concerted effort to link these bodies of knowledge and to articulate a distinct domain of knowledge at the intersection of religion and environmental politics.

Although ample social scientific scholarship examines the tremendous variegation in how climate change intersects with culture, this body of literature inadequately attends to religion, rarely raising questions about gods, institutionalized moral authorities, holy texts, or sacred places. Research on "human values" as key factors in shaping public sentiments about climate change has underscored the importance of cultural traditions, but studies of this kind seldom identify religion as a distinct cultural form for comparative analysis (e.g., Corner, Markowitz, and Pidgeon 2014; O'Brien and Wolf 2010; Weber and Stern 2011; Adger, Lorenzoni, and O'Brien 2009; Leiserowitz 2006). When religion does appear in the context of climate change and human values, it is typically operationalized as a discrete and quantifiable variable and is seldom situated in a comparative analytic framework. To take religion seriously reveals critical aspects of environmental controversy—the ethical fault lines between different stakeholder groups, the contingent histories that have allowed certain groups to access power, and the fraught necessity of moral rhetoric in climate discourse. This volume seeks to advance scholarship on the cultures of climate change and to extend them in critical new directions by attending explicitly to the power of religion in the public sphere.

Conversely, scholarship focused specifically on religion and the environment (also sometimes referred to as "religion and ecology" or "religion and nature") has a mixed record when it comes to incorporating politics in analyses. Ethical and theological methodologies are the major currents in this field of study—the textual and normative dimensions of the interface of religion and climate change, rather than politics or policy, have been most thoroughly treated (e.g., Gottlieb 2003; Posas 2007; Levett-Olson 2010). A related vein of scholarship uses quantitative methods to operationalize religion as a discrete variable within the broader political matrix, but research of this kind is not well positioned to appreciate the lived complexity of religion as a feature of contemporary public life (e.g., Morrison, Duncan, and Parton 2015; Arbuckle 2017; Ecklund et al 2017). Although there are expanding efforts among scholars of religion and the environment to bring politics directly into their analyses, these efforts remain nascent (e.g., Johnston 2014; Berry and Albro 2018). Both constructive and quantitative scholarship are essential components of any robust body of knowledge about religion and climate change; however, without methodologically rigorous attempts to collect and interpret qualitative data, it is difficult to attend to the lived experiences and emergent mobilizations of religion in contemporary environmental politics. Many key developments in religious studies have not been brought to bear in the field of religion and the environment, including the emergence of secularism studies and the expanded attention to religion's role in both domestic and international political life. Incorporating insights from these and other theoretical advances, *Climate Politics and the Power of Religion* aims to reinscribe politics as a key category for scholarship on religion and the environment.

The relationship of religion to climate politics is complex, distinguished from religious engagement in localized environmental struggles by the inherent borderlessness of the atmospheric system. Climate change has a variety of causes—greenhouse gas emissions, deforestation, agricultural intensification, aerosol and particulate pollution—and a variety of effects—tropical storm intensification, sea level rise, drought, ocean acidification, biodiversity loss, and so forth. These causes and effects are conjoined in the matrix of the earth's natural systems, but the ways in which they manifest and become tangible for human communities vary tremendously across space and time. Understanding both the religious and political aspects of this global crisis and its intimate local consequences necessitates comparison. How has the unique but ubiquitous threat of climate change created new material conditions for religious communities and new political realities for religious actors?

In seeking to address these questions, the collaborative research project from which this volume emerged was clustered around key geographic regions: the vulnerabilities of small island nations to sea-level rise and intensifying tropical storms in the Caribbean and the South Pacific; the challenges faced by high-elevation communities because of accelerating glacial melt in the Andes and Himalayas; and the experience of mounting water scarcity in large urban areas in South Asia and South America. Because this project was hosted by American University's Center for Latin American and Latino Studies, it considered how climate change will affect Latin America and how religion is mobilized in response to these impacts in other regions with similarly disrupted environmental conditions. Within this restricted frame, several dozen scholars from different nations and disciplines met regularly over a period of two years in sustained conversation to share their research. Conversations among project participants developed along three interwoven analytic streams: (a) one concerning the role of religious actors in national and transnational political contestation around climate change, (b) one concerning the role of religion and religious norms in shaping how communities experience and understand climate change, and (c) another concerning how climate change drives religious change. These three streams inform the chapters that compose this volume.

Climate Politics and the Power of Religion gathers a diverse set of empirical cases that explore the sociopolitical terrain at the intersection of religion and climate politics and is capped by a synthesis that imagines how scholarship in this domain might progress. Together, the cases presented in the various chapters advance knowledge about the political implications of religious belief and practice by asking how religious stakeholders envision and enact responses to climate change. How does religion come to matter in environmental policy debates? When, where, and on what terms is religious discourse an effective means of political intervention? What do the specific contours of climate politics indicate about the nature of religion as an organizing force in public life? How does the inclusion or exclusion of religion from systems of political power shape the possibilities for climate action? How do emerging forms of religious advocacy and mobilization around issues of climate and environment inform parallel political processes—for example, those related to transnational migration, resurgent populism, or subnational opposition to state power? Taken collectively, these questions are attempts to understand the cultural dimensions of national and global environmental politics and to appreciate the variety of ways in which religion matters in different spheres of public life.

The chapters are grouped in three parts. The first part highlights the role of religious actors and the mobilization of religious ideas in the construction of national-level climate policy. Authors in this part provide nuanced, concrete examples of how religion is marshalled in domestic politics—in each case, placing special emphasis on the limited purchase that religious actors and ideas have in different contexts. David T. Buckley tracks the uneasy relationship between the Catholic Bishops Conference and the Duterte regime in the Philippines. He argues that constructive dialogue about climate policy collapsed because of conflict over other policy issues. Presenting original ethnographic fieldwork, J. Brent Crosson unearths two competing political theologies in Trinidad that draw on opposing religious resources as they speak to the politics of domestic petroleum extraction. Neeraj Vedwan charts the contours of climate discourse in India under the Bharatiya Janata Party–led government, exploring where Hindutva ideologies are and are not salient for national environmental policy. These chapters corroborate an important finding. Even where religion might be an overt, internationally legible feature of political life—as with Hindutva movements in India or the popularity of the Catholic Church in the Philippines—its influence over climate policy carries the echoes of other contestations, including economic development, human rights, and reproductive freedom. Although climate change may be an existential issue for humanity, it does not exist apart from the dictum that "all politics is local." Across these chapters, attention to religion and religious actors is a means of revealing how climate politics are strangely handcuffed to disparate debates about crime, sexuality, ethnicity, immigration, and so forth.

The second part shifts the focus from national-level debates about climate policy to an analysis of religion as an analytic category deployed in public deliberations about climate change. Working in a transnationally comparative register, Erin K. Wilson provides a robust theoretical account of the logics of secularism, grounded in the desire for control of areas such as nature, superstition, and sexuality. Evan Berry asks about the conditions under which religious engagement in climate politics are possible and labors to shift the analytic frame from an exclusive focus on theological commitments to a broader consideration of the tactical opportunities available to religious actors. The two chapters paired in this section examine the slippery relationship between religious and secular modalities in contemporary politics, scrutinizing how these terms help construct the knowledge regimes foundational to sustainability discourse itself.

The third part articulates novel and emerging modes of religious engagement with climate politics. The chapters that compose this section illustrate the resonance and dissonance between local and international policy processes and highlight the experimental terms on which contemporary responses to climate change are put forward. Kelly D. Alley and Tarini Mehta employ legal anthropology to elicit the tensions among religious, legal, and political discourses about environmental protection in India. Ana Mariella Bacigalupo reflects on long-standing ethnographic engagement in northern Peru, articulating the shifting dynamics of cosmopolitics and showing how divine power works through local agents. Andrew R. H. Thompson incorporates theological perspectives on "vulnerability"—a term that has currency among both climate justice advocates and environmental risk analysts—to map how normative concepts circulate among the various nodes of the global climate discourse. Roger-Mark De Souza draws on extensive practitioner experience with citizen diplomacy, mapping how religion can affect the civic and transnational dimensions of climate politics. This part identifies various spaces in which religion comes to matter in climate policy and suggests why such spaces remain open and dynamic in this era of environmental precarity.

Read synoptically, *Climate Politics and the Power of Religion* develops several important contributions, evident in the resonances among chapters. Four major conclusions emerge across this volume: (a) religion is an important means by which to articulate the status of other-than-human beings in the moral economy of climate disruption; (b) climate change policy infrastructures are spaces that encourage, constrain, or refract religious engagement; (c) religion is a domain in which alternative conceptual frameworks are the basis of knowledge about environmental change; and (d) transdisciplinary theorization about religion is a useful means of comparing the social and cultural impacts of climate change. Contributors revisit and explore these topics independently, although reading the volume as a whole reveals sustained conversation among and across the chapters.

Several chapters in this volume engage conversations about the moral, legal, and political standing of other-than-human beings. The primary causes of climate disruption are anthropogenic, and many of its impacts will be acutely felt by human communities. But animals too will suffer. Rivers will dry up, and whole ecosystems—like coral reefs and alpine zones—threaten to collapse. Whether and to what extent such considerations can be folded into the anthropocentric ethical frameworks that dominate climate policy deliberations is a morally urgent question. Alley and Mehta (chap. 6) reflect directly on this issue, exploring a series of novel legal arguments in the Indian court system that assert the personhood of the Ganga River and position its human advocates as legal guardians. Attempts to legally enshrine the rights of nature generally and rivers in particular are also underway in New Zealand, Bolivia, Ecuador, and Colombia (see Bliss 2017; Albro 2018). Articulating of the interests of other-than-human beings, however, is not necessarily a juridical process. Bacigalupo (chap. 7) describes how mestizo communities in northern Peru understand climate change as the product of a system of relations among humans, *apus* (divine beings), and *wak'as* (sacred places). Environmental politics in this region, especially at the local level, may involve shamans, who are able to "channel the thoughts" of *apus* and bring such insight to bear on decision processes. Numerous environmentally disastrous state policies were made without reference to religious technologies used by many Andean peoples, and Bacigalupo concludes by addressing the how shamans (*curanderos*) present a cosmopolitical challenge to climate politics as usual. In chapter 2, Crosson advances a similar argument, drawn from an example in which informants repudiate a drilling project because it is injurious to the Earth, God's own body. These contributions are timely for ongoing scholarly engagement with the ethical, ontological, and cultural dimensions of human relations with our earthly neighbors.

A number of the contributors have considered questions about where, how, and under what conditions religious engagement with environmental politics is possible. Comparative politics, human geography, religious studies, and practitioner perspectives inform a conversation across the chapters about how the landscape within which climate change politics occurs is a limiting condition for religious actors. This argument is central in Berry's chapter 5, examining two notable cases in which religious actors were important contributors to public debates about climate change. Berry argues that such religio-political activity reflects ancillary incentives for religious leaders and institutions. That both climate change advocacy and denial can be pragmatic forms of public engagement for religious groups underscores how objectives beyond ecological sustainability remain determinative. In chapter 1, on the Catholic Bishop's Conference of the Philippines, Buckley develops the converse argument: religious actors may be reluctant to get involved with climate politics if such involvement undercuts their efficacy elsewhere. Vedwan (chap. 3) corroborates and extends these analyses, demonstrating how the application of religious symbols and ideas to public discourse does not generate a "monolithic" approach to public policy issues. Instead, his examination of Nahendra Modi's "saffron" vision of sustainability reveals how the complexity and scale of democratic institutions, civil society organizations, and religious communities in modern India act as refractories through which this vision may be realized, complicated, stymied, or transmuted. In chapter 9, De Souza focuses on one such space in civil society—the various channels of citizen diplomacy, which act as connective tissues between societies and between religious and secular institutions. In chapter 4, Wilson's scrutiny of secular policy discourse about "climate security" provides theoretical confirmation of these observations. Exploring the rhetoric of fear palpable in much of the political discourse about climate change, she identifies how secular ontologies have been spared critical scrutiny and now position global publics in the unexpectedly religious position of needing to hope for the future.

Several contributors investigate the tension between religious framings of climate change and those generated by other cultural forces. Religious discourses may sometimes prove to be sources of hope, but they do not necessarily mix easily with the dominant strains of thought about climate change—the alternative conceptual terrain on which religiously inflected climate discourse operates may spill over into or be confounded by other discursive formulations. In chapter 8, Thompson focuses on the discursive overlap regarding climate vulnerability, a keyword in both theological ethics and in policy risk assessment. He develops an account of how theological contributions might help more fully realize the category's promise as a policy diagnostic by demonstrating how efforts to build resilience follow directly from the shape and nature of ecological vulnerabilities. Vedwan (chap. 3) and Alley and Mehta (chap. 6), in their distinct approaches to contemporary environmental politics in India, make related claims. In tracing the pathways along which Hindu concepts about rivers, justice, cleanliness, and pollution travel through systems of public deliberation, these two chapters offer concrete examples of when, how, and why religious discourses do and do not find currency in policy debates. Bacigalupo's chapter 7 is one of several chapters that describe "new ways of thinking about the Anthropocene" that emerge from religious sources—in this case, the *curanderismo* that mediates relationships between landscapes and communities in northern Peru. She demonstrates a particular dimension of the power of religion: the elaboration of morally subtle, socially evocative ways of expressing human relationships with places and, perhaps more important, the possibility that these discursive and cognitive technologies can travel beyond confessional confines to resonate with broader publics. In chapter 4, Wilson supplies a helpful theoretical substrate for these intersecting analyses. She describes the dual nature of secularism, which, paradoxically, presumes the fundamental separation of human from other forms of being and demands "imaginative responses" that resecure the human position in the ecological order. This simultaneous rejection of and desire for ontologies that vivify our relations with nonhuman others serves as a powerful explanation of how religion comes to matter in environmental politics.

Running through all of the chapters is a focus on the term *religion* as a salient feature of contemporary political life. Climate change is a singular challenge that will continue to affect cultures and livelihoods everywhere and, because of its ubiquity, affords a window into the nature of twenty-first-century politics writ large. Religion can be a powerful element in the admixture of environmental politics; however, as in other arenas, this does not indicate a stable and immutable form of influence. The profundity and intensity of environmental change in the Anthropocene is a driver of religious change, and the relationship between religion and environmental politics is a dynamic one. The forms of climatic disruption, religious transformation, and political upheaval that are likely to characterize the twenty-first century merit attention as conjoined forces of change. Ken Conca's concluding chapter provides a synthesis of the cases and places described in *Climate Politics and the Power of Religion*, suggesting a manifestly postsecular world in which religion is never given as a feature of political struggles but rather is a contingent, indeterminate aspect of public life.

References

Adger, Neil, Irene Lorenzoni, and Karen O'Brien, eds. 2009. *Adapting to Climate Change: Thresholds, Values, Governance.* New York: Cambridge University Press.

Albro, Rob. 2018. "Bolivia's Indigenous Foreign Policy: Vivir Bien and Global Climate Change Ethics." In *Church, Cosmovision, and the Environment*, edited by Evan Berry and Rob Albro. London: Routledge.

Arbuckle, Matthew B. 2017. "The Interaction of Religion, Political Ideology, and Concern about Climate Change in the United States." *Society and Natural Resources* 30 (2): 177–94.

Berry, Evan, and Rob Albro. 2018. *Church, Cosmovision, and the Environment: Religion and Social Conflict in Contemporary Latin America.* London: Routledge.

Bliss, Susan. 2017. "Management: A River Is a 'Person.'" *Geography Bulletin* 49 (2): 17–18.

Corner, Adam, Erza Markowitz, and Nick Pidgeon. 2014. "Public Engagement with Climate Change: The Role of Human Values." *WIREs Climate Change* 5 (3): 411–22.

Duara, P. 2015. *The Crisis of Global Modernity: Asian Traditions and a Sustainable Future.* Cambridge, UK: Cambridge University Press.

Ecklund, Elaine Howard, Christopher P. Scheitle, Jared Peifer, and Daniel Bolger. 2017. "Examining Links between Religion, Evolution Views, and Climate Change Skepticism." *Environment and Behavior* 49 (9): 985–1006.

Gerten, Dieter, and Sigurd Bergmann, eds. 2012. *Religion in Environmental and Climate Change: Suffering, Values, Lifestyles.* London: Continuum.

Gottlieb, Roger S. 2003. *This Sacred Earth: Religion, Nature, Environment.* New York: Routledge.

Haluza-DeLay, Randolph. 2014. "Religion and Climate Change: Varieties in Viewpoints and Practices." *WIREs Climate Change* 5 (2): 261–79.

Johnston, Lucas F. 2014. *Religion and Sustainability: Social Movements and the Politics of the Environment.* Sheffield; Bristol, CT: Equinox.

Leiserowitz, Anthony. 2006. "Climate Change Risk Perception and Policy Preferences: The Role of Affect, Imagery, and Values." *Climatic Change* 77 (1): 45–72.

LeVasseur, Todd. 2015. "'The Earth Is *Sui Generis*:' Destabilizing the Climate of Our Field." *Journal of the American Academy of Religion* 83 (2): 300–319.

Levett-Olson, Laura. 2010. "Religion, Worldview and Climate Change." In *Routledge Handbook of Climate Change and Society*, edited by Constance Lever-Tracy, 261–70. New York: Routledge.

Morrison, Mark, Roderick Duncan, and Kevin Parton. 2015. "Religion Does Matter for Climate Change Attitudes and Behavior." *PLoS One* 10 (8): e0134868. doi:10.1371/journal.pone.0134868.

Norgaard, Kari. 2011. *Living in Denial: Climate Change, Emotions, and Everyday Life.* Cambridge, MA: MIT Press.

O'Brien, Karen L., and Johanna Wolf. 2010. "A Values-Based Approach to Vulnerability and Adaptation to Climate Change." *WIREs Climate Change* 1 (2): 232–42. https://doi .org/10.1002/wcc.30.

Pikhala, Panu. 2018. "Eco-Anxiety, Tragedy, and Hope: Psychological and Spiritual Dimensions of Climate Change." *Zygon* 53 (2) 545–69.

Posas, Paula. 2007. "Roles of Religion and Ethics in Addressing Climate Change." *Ethics in Science and Environmental Politics* 2007:31–49. doi: 10.3354/esep00080

Sponsel, Leslie E. 2012. *Spiritual Ecology: A Quiet Revolution.* Santa Barbara, CA: Praeger.

Weber, Elke U., and Paul C. Stern. 2011. "Public Understanding of Climate Change in the United States." *American Psychologist* 66 (4): 315–28. http://dx.doi.org/10.1037/a0023253.

Willox, Ashlee. 2012. "Climate Change as the Work of Mourning." *Ethics and the Environment* 17 (2): 137–64.

Veldman, Robin Globus, Andrew Szasz, and Randolph Haluza-DeLay. 2014. *How the World's Religions Are Responding to Climate Change: Social Scientific Investigations.* New York: Routledge.

EVAN BERRY is Assistant Professor of Environmental Humanities in the School of History, Philosophy, and Religious Studies at Arizona State University. He has previously taught at American University and Lewis and Clark College. His research examines the relationship between religion and the public sphere in contemporary societies, with special attention to the way religious ideas and organizations are mobilized in response to climate change and other global environmental challenges. Berry is the author of *Devoted to Nature: The Religious Roots of American Environmentalism* (University of California Press, 2015), which traces the influence of Christian theology on the environmental movement in the United States. Berry spent a year in residency at the State Department's Office of Religion and Global Affairs as the American Academy of Religion's inaugural Religion and International Relations Fellow. He also serves as the President of the International Society for the Study of Religion, Nature, and Culture and as Chair of the American Academy of Religion's Committee for the Public Understanding of Religion.

I.

RELIGION AND THE CONSTRUCTION OF NATIONAL CLIMATE POLICY

1

RELIGIOUS INFLUENCE AND CLIMATE POLITICS IN DUTERTE'S PHILIPPINES

Opportunity Lost?

David T. Buckley

A T FIRST GLANCE, THE PHILIPPINES MIGHT SEEM LIKE a potential poster child for the power of religious actors in shaping climate politics. Surveys indicate a highly religious population. Leaders of the Roman Catholic religious majority and several minority communities have strong local and international motivations for prioritizing climate issues in public life. In addition, there is a tradition of religious voices shaping public debates, from dramatic episodes such as the 1986 People Power Revolution to more mundane influence over areas like education policy. The Catholic Bishops Conference of the Philippines, the country's most prominent national religious body, issued a comprehensive pastoral letter on ecology, "What Is Happening to Our Beautiful Land?" in 1988 (Legaspi 1988), and environmental concerns have remained high on the agendas of a number of national and local faith-based actors. The combination of these factors with the election of a popular, tough-talking president with a history of supporting some environmental causes might make religious influence seem almost a foregone conclusion.

Nevertheless, the initial years of Philippine President Rodrigo Duterte's administration were anything but a model of smooth religious influence on environmental policy. In some areas—notably, signing the Paris Agreement—the administration lived up to the initial hopes of environmental advocates, including many religious activists. In other areas, progress has been more limited. Throughout his presidency, the potential power of religious actors to influence climate politics has been constrained by the broader patterns of friction between Duterte and various religious leaders. This friction has been spurred largely by the president's controversial "war on drugs," which has left thousands of predominantly poor Filipinos dead from police actions and shadowy vigilante executions. This finding from the Philippines comes with broader implications for the study of religious influence and climate politics: policy contention largely unrelated to environmental policy has constrained opportunities for religious influence, even in a favorable environment. This outcome suggests that religious influence over climate politics rests on other elements of policy consensus that merit attention in understanding the power of religion in environmental advocacy.

In this chapter, I document this interplay between climate politics and the influence of religious actors in the Philippines. This analysis is largely at the elite level, focused on the actions of national-level religious authorities, environmental activists, and policy makers.[1] I highlight the importance of two factors that influence religious power over environmental policy in the Philippines: (a) the institutional relationship between religion and state and (b) the broader policy priorities of religious and state elites. In this case, a permissive institutional relationship has enhanced opportunities for religious influence; however, those opportunities have been sharply constrained by misalignment between the broader policy priorities of political and religious elites. The seemingly lost opportunity of the Duterte administration illustrates a lesson with broader applicability: religious influence over climate politics is constrained or empowered by factors that may initially seem to be beyond the scope of environmental analysis.

The Religious, Political, and Environmental Landscape of the Philippines

Before moving to discuss the trajectory of religion and climate politics during the Duterte presidency, a basic introduction to the religious, political,

---

1. While beyond the scope of this chapter, scholars attempting to document forms of religious influence on climate politics would be well served by studying links between elite cues and public opinion among religious communities. Religious influence on environmental politics in the Philippines rests on strong vertical links between elites and grassroots religious society that facilitate communication and collective action. The grassroots strength of religious groups working in Indigenous communities, for instance, helps to explain the high-quality information about environmental crises facing these communities that can reach national-level religious elites.

and environmental landscape in the Philippines is in order. This brief over-view gives of a sense of the diversity within the Philippine religious land-scape, the intricate ties between religion and democracy after the 1986 People Power Revolution, the environmental challenges facing the coun-try, and the background involvement of national-level religious actors and organizations in climate politics. Although the Catholic Church predomi-nates and has since the independence period, religious minorities of vari-ous stripes are active players in the story.[2]

Analyses of the demographics of religion in the Philippines tend to begin by recognizing the predominant position of the Roman Catholic Church. Survey data from the International Social Survey Programme, carried out in conjunction with the Social Weather Stations research firm, has found Catholics at more than 80 percent of the population in each of its three waves since 1990 (Abad 2001). In contrast to many of the major-ity Roman Catholic countries in Latin America, there has been no rapid decline in Catholicism's demographic dominance.[3] Although the Catho-lic community has been the numerical center of Christian life since the consolidation of Spanish colonialism, non-Catholic Christian commun-ities also play an active role in public life. Non-Roman Catholic Christians make up approximately 10 percent of the population, a significant minority in a country of more than 100 million people.[4] This slice of the population is itself internally diverse, with denominations historically tied to Amer-ican Protestantism; nationally distinct communities such as the Philippine Independent Church and the Iglesia ni Cristo; and newer, nondenomina-tional, evangelical networks. The country also contains a sizeable Muslim

---

2. Duterte's personal biography is caught up in this complex religious landscape. He is a product of Catholic education, although he was reportedly the victim of sexual abuse at the hands of an American Jesuit priest during his youth. He famously cursed Pope Francis for causing traffic during his visit to Metro Manila and has made waves by calling God "stupid" and joking about founding a Church of Duterte. Although his own background is Catholic, Duterte developed a reputation for strong relationships with Muslim elites during his long time as a political official on the island of Mindanao.

3. While Catholicism's share of the general population has remained fairly steady, rates of religious attendance have shown some decline, and youth religiosity seems to be in a state of particular transition (Cornelio 2016).

4. Precise religious demographics are somewhat contested, particularly given the political instability in Muslim-majority portions of the Philippines. That instability almost certainly influences statistical measurement efforts. Regardless of the specific population demographics, the groups generally split with just more than 80 percent Catholic, just more than 10 percent non-Catholic Christian, and between 5 and 10 percent Muslim. See Philippine Statistics Authority (2015) for religious demographic estimates from the most recent governmental census.

minority that represents between 5 and 10 percent of the aggregate population but constitutes a majority in parts of the island of Mindanao, which is also a crucial location of many environmental controversies in the country.

At the national level, this demographic landscape is organized into a number of religious organizations that are noteworthy for this analysis because several play an active part in religious climate politics. When speaking as a body, the Catholic bishops are organized through the Catholic Bishops Council of the Philippines (CBCP), whose elected president frequently plays a prominent role in national public debates, including those related to environmental policy. In addition to the hierarchy, important Catholic voices tied to environmental politics exist in charitable organizations such as the National Secretariat for Social Action (NASSA) of the CBCP (i.e., the hierarchy's humanitarian and development arm and the national affiliate of Caritas Internationalis) and the Association of Major Religious Superiors of the Philippines, a collective of male and female religious orders. Among non-Catholic Christians, prominent organizations involved in public life include the National Council of Churches of the Philippines (NCCP), consisting of Protestant denominations like the Episcopal Church of the Philippines and the United Church of Christ in the Philippines. Many evangelical Christian churches are organized under the Philippine Council of Evangelical Churches (PCEC). Islam in the Philippines is particularly concentrated in the Bangsamoro Autonomous Region in Muslim Mindanao, with regional networks of ulama, and some national conveners such as the Philippine Center for Islam and Democracy.

The activities of these religious actors in climate politics build on intricate ties between religion and democracy in the country since the People Power Revolution of 1986 that deposed the regime of Ferdinand Marcos and restored democracy. The Marcos regime, among its crimes, broke down protections of religious liberty and patterns of cooperation between the regime and religious actors, including but not limited to the Catholic Church (Youngblood 1987, 1990). Religious networks, especially those tied to the Catholic Church, worked with election observers under the National Citizens' Movement for Free Elections to document fraud in the 1986 "snap election" that ultimately led to Marcos's fall. Manila's Cardinal Jaime Sin called for mass protest using the Catholic radio network, and religious leaders from Christian and Muslim backgrounds were active participants on the Constitutional Commission that followed Marcos's fall. Democracy in the country has remained contentious, with popular protests in 2001 deposing a subsequent president, Joseph Estrada, and corruption remaining endemic in the midst of a political system characterized by weak political parties and dominant political dynasties (Anderson 2010).

More than 25 years after the initial People Power marches, the public influence of religious actors is usually less dramatic but still substantial, resembling what political scientist Anna Grzymała-Busse (2015) has termed "institutional access." Religious actors meet regularly with government officials not only for symbolic photo opportunities but also for input on substantive policy development and engagement in institutional partnership with the state for provision of a variety of social services, from health care to disaster response. Such access is furthered through personal ties and even by clerics holding official governmental posts, such as former president Benigno "Noynoy" Aquino's secretary of education, Br. Armin Luistro, a member of the Catholic Christian Brothers. Even when religious actors "lose" policy debates, such as the Reproductive Health Act of 2012, they enjoy the opportunity for influential input into the legislative process—in that case, they secured concessions from lawmakers and then from the Supreme Court on the law's implementation and the strengthening of religious liberty protections.[5]

Climate politics attract a lot of national attention in the Philippines because the country is likely to face significant impacts from climate change and because environmental politics are closely tied to other policy challenges. These challenges include ensuring equitable economic development, responding to natural disasters, and pursuing peace between the government and long-standing insurgents tied to the political left and to Moro nationalist movements. Various studies have found the Philippines to be among the countries most vulnerable to the effects of climate change (Alave 2011; Ranada 2015; Eckstein, Kunzel, and Schafer 2017). This global position is partially tied to long-term challenges of infrastructure, rapid urbanization, and agricultural vulnerability and is likely accentuated by climate change, particularly through extreme weather events and changing precipitation patterns, as well as sea-level changes across the island group.

These environmental vulnerabilities are daunting in their own right and intersect with some of the country's most pressing public policy challenges. Governmental response to climate change is just one aspect of a

---

5. For a more detailed discussion of development of the Reproductive Health Act, with particular attention to religious influence on the legislation, see Buckley (2016) and Dionisio (2011).

broader agenda of environmental justice that extends to mining regulation, energy policy, and land management. The country has significant mineral reserves, and thus mining interests, but faces protest from local, frequently indigenous, communities affected by mining projects that have environmental costs. The country has long faced questions about its electric supply, and those question increase in the face of urbanization, economic growth, and aging existing facilities. Business, political, and environmental interests disagree about future energy sources, particularly those tied to coal and nuclear power. Several of these pressing environmental issues are centered on the island of Mindanao, where long-running insurgencies draw on a variety of social and economic grievances. The close links among environmental politics, development, and political violence help explain why religious leaders have fairly long-standing involvements in national-level policy debates that touch on climate politics.

Religious actors and organizations play regular roles in much of Philippine civil society and are closely tied to the domestic movements that shape climate-change politics.[6] The examples below are illustrative, not exhaustive. A wide variety of religious actors are involved in both elite and grassroots climate politics in the Philippines (Karaos 2011; Ramos-Llana 2011), from national-level religious figures to local "Diocesan Social Action Centers." Frequently, these religious advocates contextualize their involvement in climate politics within a broader concern for the impact of environmental issues on disaster response, poverty alleviation, and indigenous-community protection. Prominent multifaith efforts have included the Ecological Justice Interfaith Movement and convenings of Catholic, Protestant, Muslim, and Jewish leaders through the Interfaith Dialogue on Climate Change (DENR 2014). Prominent examples among the Catholic majority include the Global Catholic Climate Movement–Pilipinas and climate justice campaigns tied to religious development agencies like NASSA/Caritas and the Philippine Misereor Partnership Inc. NASSA's leader, Fr. Edwin "Edu" Gariguez, received a 2012 Goldman Environmental Prize for his environmental protection work among Indigenous communities and has remarked, "The moral imperative to act [on climate change] could not be stronger" (Gariguez 2015). The NCCP (2011)

---

6. For a comprehensive treatment of general Catholic involvement in postauthoritarian civil society movements, including some detailed analysis of local activism for environmental citizenship, see Moreno (2006).

has a history of climate advocacy, noting at its 2011 national convention, "The Earth is groaning from a climate crisis of catastrophic proportions."

This religious involvement in environmental politics includes active participation with many "secular" civil society coalitions focused on environmental politics—for instance, the Philippine Network for Climate Justice, the Green Thumb Coalition, and the 2015 March for Climate Justice Pilipinas, which coordinated a series of nationwide events in support of the 2015 United Nations Climate Change Conference in Paris. Domestic religious leadership is also tied to international climate activism. The Global Catholic Climate Movement includes a number of Philippine partners, and its Philippine affiliate launched its Laudato Si' Pledge, in support of Pope Francis's environmental encyclical, in June 2017 in Manila, with enthusiastic endorsement from Cardinal Luis Antonio Tagle (Global Catholic Climate Movement 2017). Pope Francis's unique international standing has buttressed local attention to religion and the environment, particularly through his emotional visit in the aftermath of Typhoon Haiyan (locally called Typhoon Yolanda) and consistent insistence that climate change remain a moral priority, especially because of its impact on the global poor. Within evangelical Protestantism, the former head of the PCEC, Bishop Efraim Tendero, has served as Secretary General and CEO of the World Evangelical Alliance since 2015. In that role, he has helped to lead global evangelical networks engaged with negotiations tied to the Paris Agreement (World Evangelical Alliance 2015). Filipino religious elites both shape and are shaped by international climate politics.

A brief examination of the involvement of religious actors in the response to Typhoon Haiyan, the country's greatest recent natural disaster, clarifies the diverse pathways through which religious groups engage with climate politics and the intersection of climate politics with a broad range of policy challenges facing the country (Wilkinson 2017). The typhoon struck in early November 2013, affecting a vast swath of the middle of the country, centered on the city of Tacloban. More than seven thousand people died, and hundreds of thousands required immediate and lasting assistance. Religious organizations were at the forefront of both immediate disaster response and longer-term efforts to address the impact of the storm. Multinational religious aid networks like Caritas, World Vision, and agencies affiliated with the ACT Alliance were at the front lines of disaster response efforts, in partnership with local community and faith-based organizations. This assistance included providing short-term essentials like medicine and food and meeting longer-term needs tied to reconstruction of schools, construction of housing, and legal assistance. As a part of long-term recovery, religious response networks became involved in policy decisions about building codes, land use, and agricultural strategy. And the same religious actors providing first-response emergency aid were drawn into policy responses to climate change at the national and even international level. As Michel Roy, secretary general of Caritas Internationalis (2015), noted on Haiyan's two-year anniversary in 2015, "The basic injustice remains—every day we wait to come to a binding global climate agreement, more and more people die in the world's low-income countries."

As Haiyan showed in tragic relief, religious actors are at the center of environmental politics in the Philippines in a variety of ways. The intersection of a vibrant religious landscape with complex environmental challenges has presented ample opportunity for religious dynamics to intersect with climate politics in the Philippines, making communities more environmentally resilient and shaping international treaty negotiations. It is a key reason why a new presidential administration prioritizing environmental concerns could have found supporters in the religious community.

The Importance of Institutions: "Benevolent Secularism" and Climate Politics

The potential opportunity for religious actors to shape national-level environmental debates owes a great deal to the nature of the institutional relationship between religion and state in the Philippines. Although this institutional relationship is not sufficient to ensure religious influence, it greatly increases the opportunities available to religious actors. In contrast with stricter forms of religion–state separation, the Philippines has instituted a softer form of secularism, which I have termed "benevolent secularism."[7] One would expect that different political institutions in other countries, particularly any that restricted religious autonomy, would have quite different effects on the opportunities for religious influence.

Although Article II, Section 6, of the Philippine Constitution declares, "The separation of Church and State shall be inviolable," the institutional

---

7. This phrasing borrows from a Philippine Supreme Court case that dubs the religion-state relationship as "benevolent neutrality," *Estrada v. Escritor*, Supreme Court of the Philippines (2003). See Buckley (2017) for a more expanded treatment of the origins and dimensions of this institutional relationship.

secularism in the 1987 Constitution was never understood in anticlerical terms. Rather, that document and its later interpretation by the courts and in the day-to-day practice of politics has developed into a form of secularism that explicitly encourages religious involvement in public life, including the development of public policy. Basic differentiation of religion and state combines with extensive cooperation between religious and state institutions. Hilario Davide, future chief justice of the Supreme Court, remarked during the 1986 Constitutional Commission's deliberations, "The new Constitution must be pro-God, pro-people, pro-Filipino."[8] And cooperation between religion and state is not simply a matter of the Catholic Church; in echoes of what Rajeev Bhargava (2007) has termed "principled distance" between the state and religious communities, the Philippine state actively cultivates cooperative relations with a variety of religious actors, particularly from Christian and Muslim communities. This benevolent form of secularism is apparent in several areas, including education policy. Article XIV, Section 4, exempts religious schools from requirements of Filipino ownership, in recognition of the "tremendous contribution to the country" that Constitutional Commissioner Rama described these schools making to Filipino education.[9]

The nature of benevolent secularism encouraged religious involvement in climate politics in the years before the Duterte administration came into power. Institutionalized patterns of consultation between state and religious actors were evident in the 2014 Interfaith Dialogue on Climate Change, which blended contributions (and funding) from religious, governmental, and international organizations (Lores 2014). While Catholic leaders were certainly prominent, the consultation also included various Protestant and Muslim elites. Access for religious elites increases when key policy makers have personal and professional ties to religious institutions. As an example, Tony La Viña, previously the lead negotiator for the Philippines in international climate-change summits, has also served in several posts in universities tied to the Philippine Jesuits. Benevolent secularism encourages influence to extend from consultation to programmatic cooperation. In the immediate aftermath of the People Power Revolution, for instance, grassroots Catholic Basic Ecclesial Communities convinced the

---

8. Constitutional Commission of the Philippines, *Record of the Constitutional Commission: Proceedings and Debates*, 5 vols. (Quezon City, Metro Manila: Constitutional Commission, 1986), 4:368.

9. Constitutional Commission, *Record*, 368.

government to ban a logging project and then "entered into partnership with the Department of Energy and Natural Resources in implementing the log ban and in the reforestation program" (Picardal 2015). This pattern of influence at times brought religious activists into conflict with political authorities. In 2015, Sr. Cres Lucero, convener of Philippine Misereor Partnership Inc (a church development agency), criticized the Aquino administration for "forget[ing] its responsibility to take the lead in protecting the Philippines' rich biodiversity and natural resources" (Samson 2015) when his State of the Nation Address neglected to raise environmental protection questions related to mining.

This institutional context, a decidedly benevolent form of secularism that encourages broad involvement of religious actors in the public policy process, including cooperation between religion and state in developing and implementing policy, is crucial to understanding the opportunity that seemed open for religious actors to shape climate politics in the Philippines. Lawmakers are accustomed to consultations with national-level religious elites, and those elites have ample experience navigating the Philippine policy process.

Climate Politics and a Duterte Opportunity?

As then-mayor Rodrigo Duterte surged in public support in advance of the May 2016 elections, both national and international media tended to focus on his persona as a tough-talking official from outside of the traditional Manila power elite, with a record of popularity in Davao City while being dogged by allegations of ties to human rights abuses. Although environmental policy received less mainstream attention, interested observers noted a record that seemed to present some opportunities for environmental progress—notably, in skepticism toward mining interests and defense of Indigenous communities affected by aspects of industrialization. There were questions about how his tough talk would translate into tangible policy. However, climate politics seemed to provide an area in which the incoming president could find common ground with civil society, including many religious elites.

In some ways, the early months of the Duterte administration seemed to bear out these hopes. His initial State of the Nation Address (commonly referred to as the "SONA") in July 2016 prominently mentioned environmental concerns, notably "illegal logging, illegal mining . . . and other destructive practices that aggravate the devastation of our natural resources" (Duterte 2016). The address went on to discuss the challenges of watershed destruction, land use, pollution, agricultural and fisheries policy, and links between environmental protection and ensuring "a human approach to development and governance." Duterte voiced support for a National Land Use Act designed to assist with climate-change mitigation and adaptation through responsible planning policy (Marin 2015). Most notably on the international stage, the president eventually signed the Philippines onto the Paris Agreement in March 2017, with the agreement taking effect a month later.

The most tangible steps translating this pro-environment rhetoric into action came as Duterte nominated a noted environmental activist, Gina Lopez, to be his secretary of the DENR. The DENR plays an important part in environmental regulation, particularly related to mining, and Lopez was widely seen as a nominee who would be less favorable to mining interests than typical cabinet members. She quickly initiated a compliance audit of mining companies, which led to a wide range of suspensions and cancelations (Geronimo 2017). Under her leadership, the DENR also announced broad restrictions on open-pit mining. In his 2016 SONA, Duterte acknowledged that "many are complaining" about Lopez's appointment but seemed to offer public support for her actions in front of the assembled lawmakers and public officials.

This early record on environmental issues was welcomed by many environmental groups affiliated with religious communities. As Duterte's inauguration approached in June 2016, Catholic leaders gathered on the first anniversary of Pope Francis's encyclical Laudato Si' to call on the new administration to prioritize environmental concerns in its early days in office (Remitio 2016). Addressing a potential crackdown on mining abuses, Catholic Archbishop Ramon Arguelles of Lipa placed particular emphasis on Duterte's background: "I know he means it . . . [the incoming President knows] the Mindanao experience" (*Manila Times* 2016). Among Duterte's most vocal religious optimists was Fr. Joel Tabora, SJ, president of Ateneo de Davao University, a prominent Jesuit university in Duterte's political base of Davao City. In the early days of the administration, Tabora (2016) pointed to environmental issues in particular as a source of his optimism, stating, "So one of the things that have been very important for our people here is his commitment against large scale mining." Lopez's appointment attracted significant praise, with Arguelles stating, "We thank

Figure 1.1. *From left to right*: Tabora, Duterte, Valles, and Lopez. Public domain.

God for her dedication to bring change for the better in the graft-ridden department that has not honestly served our people particularly the poor" (CBCP News 2017). Tabora (2017b) was even more effusive on Twitter after the Lopez appointment, remarking, "Finally, a Dept of ENVIRONMENT and NATURAL RESOURCES, not a Dept of EXPLOITATION and ALIEN-ATED RESOURCES! #ConfirmSecGina." Duterte appeared with Lopez, Fr. Tabora, and Archbishop Romullo Valles of Davao City at an August 2016 summit entitled, "OYA Mindanaw! State of Mindanao Environment Day" (fig. 1.1).

It is important to note that from the beginning, Duterte's relationship to climate politics was more ambiguous than some in the environmental movement would have preferred. In his 2016 SONA, given before his signing of the Paris Agreement, Duterte clearly voiced skepticism about the Philippines limiting industrialization to address a problem caused by more significant emissions from global powers like the United States, China, and the European states. He has supported new coal-fired power plants, which make up a large share of the country's strained power generation, and made ambiguous statements about re-emphasizing the country's nuclear power sector, centered on the Bataan Nuclear Power Plant. Most prominently, despite his public support for Gina Lopez (his nominee to head the DENR), the country's powerful Commission on Appointments rejected her appointment after questioning her early regulatory decisions.[10] Duterte appointed a former military official in her place, and regulatory moves against mining interests have been slowed or rolled back. In his 2017 SONA, Duterte "sternly warn[ed]" the mining sector to abide by regulations and insisted that "protection of the environment must be made a priority" (Duterte 2017).[11] One environmentalist group, however, expressed doubts about the "cheap talk" of the address in light of Gina Lopez's rapid fall from the administration (Gamil 2017).

A Source of Missed Opportunity: Constraint from Clashing Policy Priorities

Despite several factors seeming to set the stage for religious influence over climate politics in the Philippines, policy progress has been uneven. Even where progress has taken place, it is not clear that religious influence has been decisive. What accounts for this missed opportunity? The proximate answer is that both political and religious actors have found cooperation on environmental policy to be constrained because of the simmering discontent about other policy priorities of the Duterte administration. This controversy includes the brutal conduct of Duterte's "war on drugs" but extends to unease about his treatment of the Marcos dictatorship, declaration of martial law across Mindanao, and heightened prospects of restrictions on free speech and liberty of the press. In a lesson that holds broader resonance for the study of religious power and climate politics, a contentious policy agenda largely unrelated to environmental concerns has substantially constrained avenues for religious influence, even in a highly favorable overall environment.

Relations between Duterte and religious elites were mixed even before his comfortable victory in the 2016 election. As noted earlier, some prominent religious leaders, especially those who worked with him during his time as mayor of Davao City, expressed optimism about his election.

---

10. The Commission on Appointments has broad authority to confirm or deny cabinet nominees in the Philippines. However, given the high degree of authority in the Philippine executive branch, presidents tend to have their nominees approved by the body. Lopez was one of several Duterte nominees rejected by the commission, in a highly unusual pattern.

11. Duterte's 2018 SONA featured a similar pattern, with specific remarks on the need for disaster preparedness, mining regulation, and "intergenerational responsibility" to care for environmental resources (Duterte 2018).

He is reported to have strong relationships with local Muslim elites and performed extremely well in Muslim-majority portions of Mindanao on election day. However, even before the election, elements of his tough-on-crime populist persona raised serious concerns among many religious leaders, particularly some in the Catholic hierarchy. Archbishop Antonio Ledesma, who serves in Cagayan de Oro (one of Mindanao's largest cities outside of Duterte's Davao), decried the mayor's lack of action to address roughly fourteen hundred vigilante killings tied to a shadowy group known as the "Davao Death Squad" as contributing to a "culture of death" in the city (Ledesma 2016). Referring to these allegations and to Duterte's comments about women, Archbishop Socrates Villegas (2016), then-president of the CBCP, wrote in the days before the election that "a choice for a candidate who takes positions that are not only politically precarious but worse, morally reprehensible, cannot and should not be made by the Catholic faithful and those who take their allegiance to Christ and his Kingship seriously."

There was little improvement in the relationship between Catholic elites and the Duterte administration during his first years in office. The policy priorities of the new administration were areas that provoked conflict, not cooperation, and thus constrained the possibility of cooperation in pursuit of environmental policy goals. There is no doubt that the most controversial policy is the administration's approach to the war on drugs, which has left thousands dead and provoked searing criticism from many of the same religious leaders who could have served as allies in environmental campaigns. As a brief overview, Duterte made his name in Davao City in large part through coercive antidrug policies, notably taking little action to stop the vigilante Davao Death Squad's executions of alleged drug users and dealers. He has boasted of taking part in killing several crime suspects. On the campaign trail, he pledged to kill one hundred thousand criminals if necessary and to fatten the fish of Manila Bay with their corpses.

On assuming the presidency, this antidrug strategy moved to the national stage, with brutal results. Estimates of total deaths from police operations and extrajudicial killings vary, but Human Rights Watch (2018) reports that credible counts had already approached twelve thousand by the close of 2017, and the acronym *EJK* (for *extrajudicial killing*) has entered everyday vocabulary. This approach has raised a range of international and domestic protest, notably from faith leaders who are especially concerned with the protection of human life and the disproportionate impact of extrajudicial killings on the urban poor (Brooke, Buckley et al, Forthcoming). Although it took the CBCP some time to find its unified voice in opposition to the violent approach to antidrug operations, it declared without reservation in September 2017, "In the name of God, stop the killings! May the justice of God come upon those responsible for the killings!" (Villegas 2017). Elites from the NCCP have expressed similar concerns, with some clergy receiving threats in response. As the NCCP's general secretary put it, "Church people who stand by the rights of the people are being called communists" (Olea 2019).[12]

There is no doubt that the war on drugs is the primary policy priority that has constrained religious influence on environmental politics; however, the sources of disagreement are more diverse. A second set of tensions could be loosely grouped together as threats to civil liberties and democratic consolidation. As mentioned, religious actors were central to the largely peaceful People Power Revolution that restored democratic institutions after the descent into martial law and dictatorship under Ferdinand Marcos. Many religious actors view safeguarding the legacy of People Power and avoiding a return to authoritarianism as among their highest priorities in public life. Duterte has provoked concerns in several ways. He has allowed Marcos's corpse to be interred in the country's national heroes memorial despite protests from victims of human rights abuses under martial law. He declared (and extended) martial law throughout Mindanao in response to the terrorist takeover of Marawi City in May 2017. He has supported reintroducing the death penalty in the country. He has publicly rebuked critical media outlets by name, and his supporters in the Congress refused to renew the broadcast franchise of one of the country's largest radio and television networks. Allies in the House of Representatives voted to defund the constitutionally mandated Commission on Human Rights, which has been highly critical of the conduct of the war on drugs, before eventually funding the agency at a reduced level after national and international outcry. Duterte has advocated for amendments to the 1987 Constitution that could introduce needed devolution of authority to regional governments but also could subvert rights protections enshrined in that document. Local elections were delayed, and the president's leading critic

---

12. It is important to note that ethnographic research indicates that some grassroots religious clergy, particularly from evangelical backgrounds, are enthusiastic supporters of the drug war (see Cornelio and Medina 2019).

in the Senate imprisoned on charges that many suspect are politically motivated. Allies forced the removal of the Supreme Court's chief justice, who had questioned aspects of the drug war, and Duterte himself undermined the authority of the country's ombudsman, charged with investigating corruption.[13]

These policy priorities have strained relationships with precisely the environmental groups and religious actors that might have been allies in climate politics. Environmental groups like the KALIKASAN network are not only frustrated with the rollback of mining reviews launched by Gina Lopez but also broadly critical of human rights violations and press restrictions attributed to the Duterte administration. This is in part because alleged human rights violations affect not only those tied to the war on drugs but also environmental and Indigenous activists. The Philippines has long been a dangerous place to engage in local environmental protest, but activists argue that "this has become even more pronounced under the current regime of President Rodrigo Duterte. In just over a year under Duterte, at least 33 environmental defenders have been killed" (Dulce 2017).[14] Understanding the deteriorating ties between religious environmentalists and the administration requires attention to the policy context of environmental debates—in this case, the extent to which those debates are linked directly to controversies about human rights protections. Environmental coalitions have been active in protests against alleged human rights violations in Duterte's Philippines.

A similar trend has played out among leading religious voices who are sympathetic to environmental advocacy but find their ability to influence governmental policy constrained by the broader policy environment. The Ecological Justice Interfaith Movement network has publicly criticized not only the treatment of Gina Lopez by the Commission on Appointments but also the rise in extrajudicial killings and the Duterte administration's proposals to restore the death penalty. Archbishop Rolando

---

13. For a comprehensive overview of concerns with diverse aspects of the Duterte administration's agenda from a social justice research center at the country's most prominent Jesuit university, see JJCICSI (n.d.).

14. Even local clergy may have become a part of this statistic. In December 2017, gunmen assassinated Fr. Marcelito Paez, and in May 2018, attackers murdered Fr. Mark Ventura immediately after Mass. The motives for the assassinations have not been definitively established, but both priests were active in environmental justice and social development advocacy in their respective dioceses. Duterte has publicly alleged that Fr. Ventura's murder was due to affairs with married women, further inflaming tensions with religious authorities.

J. Tirona, then director of NASSA/Caritas Philippines, stated that restriction of media outlets "smacks of 'reign of terror,' an arrogant and high-handed warning to the media practitioners that do not kowtow to the administration" (Rappler.com 2018). Fr. Edu Gariguez, NAASA/Caritas's executive director and a winner of the Goldman Environmental Prize, recently said that "the authoritarian rule of Duterte is beginning to become even worse than the martial law of Marcos, which he tried to disguise through some semblance [of] legality. Duterte is more brazen, unreasonably vindictive, with little or no regard for accountability" (Williams 2018). Other faith groups involved in environmental advocacy, including the NCCP and the PCEC, have voiced public concerns about extrajudicial killings in particular. As the PCEC's leadership wrote in an August 2017 statement, "We denounce in the strongest terms all unjust slayings, and the rising culture of senseless killing and impunity that, to our very deep regret, have accompanied from its very inception the President's well-intentioned campaign against illegal drugs" (Pantoja 2018). Even Duterte's clerical defenders, who often justified their support with references to his environmental views, have found their position more difficult to maintain. Fr. Joel Tabora issued a letter in early 2017 stating, "If I must choose between going to hell with President Duterte in pursuit of the war on drugs in the Philippines or going to heaven with Abp. Soc Villegas because neither he nor any of the Catholic Bishops of the Philippines 'find pleasure in the death of anyone who dies' (cf. Ezekiel 18:32), I choose going to heaven with the CBCP" (Tabora 2017*a*).

In summary, the potential for religious dynamics to shape climate politics in the Philippines under the presidency of Rodrigo Duterte has been substantially constrained by the broader policy priorities of both the president and key religious actors. This is not only, or even primarily, because the president and those actors disagree on the substance of climate politics. Indeed, in signing the Paris Agreement and maintaining at least rhetorical pressure on mining interests, raw material remains for religious advocates to work with in the administration's policy approach. And the institutions of benevolent secularism in the Philippines make it a favorable environment for such influence to take place. Rather, despite these opportunities, religious influence over climate politics has been constrained because of tensions regarding the broader policy priorities of key actors, especially those tied to the war on drugs, and broader concerns about Duterte's impact on the stability of democratic institutions. In the immediate future, religious influence on climate politics is another piece of collateral damage in the war on drugs.

Implications: Religion and Climate Politics beyond the Philippines

The recent history of religious power and climate politics in the Philippines provides several lessons that could inform the study of religion and the environment in comparative contexts. Although national settings will vary, the opportunity lost in Duterte's Philippines has several implications for the study of religion and climate politics beyond this case.

First, the institutional relationship between religion and state is a crucial, if not determinate, factor structuring the influence of religious actors on climate politics. Political scientists have paid increasing attention to varieties of legal relationships between religion and state, even among democracies (Stepan 2000; Fox 2006; Kuru 2009; Buckley 2015). A general consensus has emerged that, counter to the assumption of some liberal theorists, democracy does not require the separation of religion and state to thrive—in fact, a wide variety of institutional alternatives to strict separation exist on the global stage (Cady and Hurd 2010; Calhoun, Juergensmeyer, and VanAntwerpen 2011). These institutional arrangements matter for religious politics, including the treatment of minority groups, the role of religious institutions in social welfare provision, and the place of religion in foreign policy. The evidence from the Philippines indicates that this institutional relationship also shapes the nature of religious influence in climate politics. The "benevolent secularism" that has characterized Philippine institutions since the People Power Revolution provides significant input for religious actors in the policy process, including consulting with legislators and elected officials on the development of climate legislation, working alongside negotiating teams at global forums like the COP talks, and entering into programmatic partnerships with government agencies in implementing climate adaptation programs.

It is important to remember that political institutions may both constrain and facilitate religious involvement. This is the case in more authoritarian settings, where a range of restrictions on religious actors may constrain their ability to exercise independent influence over environmental politics. Government restrictions may make religious actors hesitant to criticize damaging environmental policies or even reduce the capacity of religious actors to organize and mobilize in public life. The comparative lesson is that institutions matter, and scholars looking to make sense of the impact of religious actors on climate politics would be well served by understanding the institutional history of the religion–state relationship in a given national setting.

Second, religious influence on environmental politics takes place within a complex set of policy priorities, on the parts of both political and religious elites, that may constrain influence even where other factors seem conducive to religious input. In the Philippines, this influence has largely been seen through the crisis brought on by President Duterte's approach to the war on drugs. From his earliest days in office, this policy has been the top priority for the president and his allies. It has been carried out with such brutality that even potential religious allies have found it increasingly impossible to simply look the other way and cooperate on other policy priorities where they share common ground. The war on drugs has turned an environment that seems ripe for religious influence into an opportunity that looks to be largely lost, at least at the national level.

President Duterte has a unique set of policy priorities, and it is unlikely that other cases will experience identical sources of polarization. However, it is not difficult to summon anecdotal evidence of other clashing policy priorities limiting potential religious influence over climate politics. In the United States, conflicts between the Obama administration and certain religious actors regarding perceived violations of religious liberty in the Affordable Care Act posed a persistent challenge to the administration's relationship with some religious elites. Perhaps this conflict limited opportunities for collaboration on environmental issues with religious actors who were sympathetic to the Paris Agreement. The sources of tension could vary widely. The important reminder is that religious influence over climate politics will take place within a broader set of policy priorities that shape potential for that influence.

The Philippine case highlights the constraint imposed by conflicting policy priorities. Nevertheless, it is entirely possible that broader policy priorities may facilitate religious influence over climate-change debates in cases that initially seem less likely for such input. In such cases, religious influence over climate politics rests on other elements of policy consensus, which then enhance religious influence on unrelated climate policy debates. As an example from public health, significant evidence suggests that evangelical Christian activists played an important part in the launch of the President's Emergency Program for AIDS Relief (PEPFAR) by the George W. Bush administration (Busby 2010). These Christian activists found particular influence over health policy because of the alignment of their broader policy priorities with those of the "compassionate conservatism" of the early years of the Bush administration. In this instance, religious framing and advocacy of public health policy drove policy change because of shared priorities around a form of Christian humanitarianism and a desire to expand cooperation between the government and religious charitable organizations.

Philippine political institutions are facing a period of great uncertainty. What one can say with relatively high confidence is that the challenges of climate politics are unlikely to recede from the public agenda, as storm intensification, sea-level rise, and shifting climate patterns affect spheres from urban planning to rural agriculture. Political and religious elites are sure to maintain interests in addressing these challenges. Many basic conditions that make the country a candidate for significant religious influence remain present. But it may take a divine intervention to overcome the policy obstacles dividing religious environmental activists from the current administration.

References

Abad, Ricardo G. 2001. "Religion in the Philippines." *Philippine Studies* 49 (3): 337–55.

Alave, Kristine. 2011. "Philippines Ranks Third on Climate Change Vulnerability List." *Philippine Inquirer*, October 10, 2011.

Anderson, Benedict. 2010. *Cacique Democracy in the Philippines: Origins and Dreams*. London: Routledge.

Bhargava, Rajeev. 2007. "The Distinctiveness of Indian Secularism." In *The Future of Secularism*, edited by T. N. Srinivasan, 20–53. Oxford: Oxford University Press.

Brooke, Steven, David T. Buckley, Clarissa David, and Ronald Mendoza. "Religious Protection from Populist Violence: The Catholic Church and the Philippine Drug War." *American Journal of Political Science*. Forthcoming.

Buckley, David T. 2015. "Beyond the Secularism Trap: Religion, Political Institutions, and Democratic Commitments." *Comparative Politics* 47 (4): 439–458.

———. 2016. "Demanding the Divine? Explaining Cross-national Support for Clerical Control of Politics." *Comparative Political Studies*. doi:10.1177/0010414015617964.

———. 2017. *Faithful to Secularism: The Religious Politics of Democracy in Ireland, Senegal, and the Philippines*. Religion, Culture, and Public Life. New York: Columbia University Press.

Busby, Joshua W. 2010. *Moral Movements and Foreign Policy*. Cambridge Studies in International Relations. New York: Cambridge University Press.

Cady, Linell Elizabeth, and Elizabeth Shakman Hurd. 2010. *Comparative Secularisms in a Global Age.* Houndmills, UK: Palgrave Macmillan.

Calhoun, Craig J., Mark Juergensmeyer, and Jonathan VanAntwerpen. 2011. *Rethinking Secularism.* New York: Oxford University Press.

Caritas Internationalis. 2015. "Typhoon Haiyan: Two Years on the Climate Isn't Changing." Caritas Internationalis, November 5, 2015. Accessed January 26, 2018. https://www .caritas.org/2015/11/typhoon-haiyan-two-years-on-the-climate-isnt-changing/.

CBCP News. 2017. "Keep DENR Chief Lopez, Archbishop Urges Duterte." CBCP News, February 1, 2017. Accessed January 23, 2018. https://cbcpnews.net/cbcpnews/keep -denr-chief-lopez-archbishop-urges-duterte/.

Cornelio, Jayeel, and Erron Medina. 2019. "Christianity and Duterte's War on Drugs in the Philippines." *Politics, Religion, and Ideology* 20 (2): 151–69. doi:10.1080/21567689.2019.1 617135.

Cornelio, Jayeel Serrano. 2016. *Being Catholic in the Contemporary Philippines: Young People Reinterpreting Religion.* Religion in Contemporary Asia. London: Routledge, Taylor & Francis Group.

DENR (Department of Environment and Natural Resources). 2014. "Interfaith Dialogue Links Religion with Climate Change." DENR, November 2014. Accessed January 24, 2018. http://r5.denr.gov.ph/index.php/86-region-news-items/469-interfaith-dialogue -links-religion-with-climate-change.

Dionisio, Eleanor R. 2011. *Becoming a Church of the Poor: Philippine Catholicism after the Second Plenary Council.* Quezon City, Philippines: John J. Carroll Institute on Church and Social Issues.

Dulce, Leon. 2017. "War in Paradise in the Time of Duterte." Rappler.com, December 11, 2017. Accessed January 26, 2018. https://www.rappler.com/views/imho/190766-duterte-war -paradise-environment.

Duterte, Rodrigo. 2016. "Transcript: State of the Nation Address." Rappler.com, July 26, 2016. Accessed January 22, 2018. https://www.rappler.com/nation/140860-rodrigo-duterte -speech-sona-2016-philippines-full-text.

———. 2017. "Transcript: State of the Nation Address 2017." Rappler.com, July 25, 2017. Accessed January 22, 2018. https://www.rappler.com/nation/176566-full-text-president -rodrigo-duterte-sona-2017-philippines.

———. 2018. "Transcript: State of the Nation Address 2018." Rappler.com, July 23, 2018. Accessed June 16, 2021. https://www.rappler.com/nation/rodrigo-duterte-sona-2018 -philippines-speech.

Eckstein, David, Vera Kunzel, and Laura Schafer. 2017. *Global Climate Risk Index 2018*, edited by Germanwatch. Bonn: Germanwatch.

Fox, J. 2006. "World Separation of Religion and State into the 21st Century." *Comparative Political Studies* 39 (5): 537–69.

Gamil, Jaymee. 2017. "Environmental Activists on Duterte SONA: Less Talk, More Action." *Philippine Daily Inquirer,* July 25, 2017.

Gariguez, Edwin. 2015. "We Have a Moral Imperative to Act on Climate Change." *Inter-Press Service News Agency,* June 17, 2015.

Geronimo, Jee. 2017. "DENR Announces Closure of 23 Mining Operations." Rappler.com, February 5, 2017. Accessed January 22, 2018. https://www.rappler.com/nation/160270 -denr-closes-mining-operations.

Global Catholic Climate Movement. 2017. "Laudato Si' Pledge Launched on 2nd Anniversary of Encyclical's Release." *Medium*, June 29, 2017. Accessed January 23, 2018. https:// medium.com/@CathClimateMvmt/laudato-si-pledge-launched-on-2nd-anniversary -of-encyclical-s-release-9d5c410ee527.

Grzymała-Busse, Anna. 2015. *Nations under God: How Churches Use Moral Authority to Influence Policy*. Princeton, NJ: Princeton University Press.

Human Rights Watch. 2018. "Philippines: Duterte's 'Drug War' Claims 12,000+ Lives." Human Rights Watch, January 18, 2018. Accessed January 24, 2018. https://www.hrw .org/news/2018/01/18/philippines-dutertes-drug-war-claims-12000-lives.

JJCICSI (John J. Carroll Institute on Church and Social Issues). n.d. "Lights and Shadows." Accessed June 16, 2021. http://www.jjcicsi.org.ph/lights-and-shadows/.

Karaos, Anna Marie A. 2011. "The Church and the Environment: Prophets against the Mines." In *Becoming a Church of the Poor: Philippine Catholicism after the Second Plenary Council*, edited by Eleanor Dionisio, 53–60. Quezon City, Philippines: John J. Carroll Institute on Church and Social Issues.

Kuru, Ahmet T. 2009. *Secularism and State Policies toward Religion: The United States, France, and Turkey*. Cambridge, UK: Cambridge University Press.

Ledesma, Antonio. 2016. "Pastoral Letter: A Matter of Conscience." MindaNews.com, May 8, 2016. Accessed January 24, 2018. http://www.mindanews.com/mindaviews /2016/05/pastoral-letter-a-matter-of-conscience/.

Legaspi, Leonardo. 1988. "What Is Happening to Our Beautiful Land?" Catholic Bishops Conference of the Philippines, January 29, 1988. Accessed January 26, 2018. http:// www.cbcponline.net/documents/1980s/1988-ecology.html.

Lores, Rex. 2014. *Interfaith Dialogue on Climate Change 2014: Our Choice*. Manila: Climate Change Commission of the Philippines.

Manila Times. 2016. "CBCP Calls for Repeal of Mining Act." *Manila Times*, June 12, 2016.

Marin, Gemma Rita. 2015. *Land Use Planning for Mitigation and Adaptation to Climate Change Now*. Quezon City, Philippines: Campaign for Land Use Policy.

Moreno, Antonio F. 2006. *Church, State, and Civil Society in Postauthoritarian Philippines: Narratives of Engaged Citizenship*. Quezon City, Philippines: Ateneo de Manila University Press.

NCCP (National Council of Churches in the Philippines). 2011. "On Environmental Protection and Climate Change Adaptation." NCCPhilippines.org, December 5, 2011. Accessed January 24, 2018. http://nccphilippines.org/2011/12/05/on-environmental -protection-and-climate-change-adaptation/.

Olea, Ronalyn V. 2019. "Church Leaders Call on Public to Defend Human Rights, Pursue Justice." *Bulatlat*, January 22, 2019. Accessed June 16, 2021. https://www.bulatlat .com/2019/01/22/church-leaders-call-on-public-to-defend-human-rights-pursue -justice/.

Pantoja, Noel Alba. 2018. "PCEC State on the Recent Upsurge of Drug-Related Killings." Philippine Council of Evangelical Churches, August 25, 2017. Accessed January 26, 2018. http://pcec.org.ph/2017/08/25/pcec-statement-on-the-recent-upsurge-of-drug -related-killings/.

Philippine Statistics Authority. 2015. *2015 Philippine Statistical Yearbook*. Manila: Philippine Statistics Authority.

Picardal, Amado. 2015. "The Philippine Church's Environmental Advocacy." Pahayagang Balikas. Accessed June 16, 2021. https://issuu.com/balikasonline/docs /vol._20__no._29_-_july_20_-_26__201.

Ramos-Llana, Melanie. 2011. "The Church of the Poor in the Province of Albay: The Diocesan Social Action Center of Legazpi." In *Becoming a Church of the Poor: Philippine Catholicism after the Second Plenary Council*, edited by Eleanor Dionisio, 61–72. Quezon City, Philippines: John J. Carroll Institute on Church and Social Issues.

Ranada, Pia. 2015. "PH Drops in 2016 List of Countries Vulnerable to Climate Change." Rappler.com, November 18, 2015. Accessed January 19, 2018. https://www.rappler.com /science-nature/environment/113064-philippines-2016-climate-change-vulnerability -index.

Rappler.com. 2018. "Caritas Philippines: 'A Government That Curtails Press Freedom Is Insecure.'" Rappler.com, January 24, 2018. Accessed January 24, 2018. https://www .rappler.com/moveph/caritas-philippines-press-freedom-insecure-government.

Remitio, Rex. 2016. "Duterte Urged to Prioritize Climate Change Action." CNN Philippines, June 18, 2016. Accessed January 26, 2018. http://cnnphilippines.com/news/2016/06/18 /Duterte-Pope-Francis-climate-change-action.html.

Samson, Oliver. 2015. "Aquino Overlooked Environmental Issues in SONA—Group." CBCPNews, July 31, 2015. Accessed June 16, 2021. https://cbcponlineradio .com/aquino-overlooked-environmental-issues-in-sona-group/.

Stepan, A. 2000. "Religion, Democracy, and the 'Twin Tolerations.'" *Journal of Democracy* 11 (4): 37–57.

Tabora, Fr. Joel. 2016. "Interview with Ateneo de Davao's Fr. Tabora on Duterte." Inquirer. net, June 28, 2016. Accessed January 22, 2018. http://newsinfo.inquirer.net/792859 /full-text-interview-with-ateneo-de-davaos-fr-joel-tabora-on-duterte.

———. 2017a. "Choosing between Hell with Duterte and Heaven with the CBCP." *Fr. Joel E. Tabora, S.J.* (blog), February 11, 2017. Accessed January 26, 2018. https://taborasj .wordpress.com/2017/02/11/choosing-between-hell-with-duterte-and-heaven-with -the-cbcp/.

——— (@Joeltaborasj). 2017b. "Finally, a Dept of ENVIRONMENT and NATURAL RESOURCES, not a Dept of EXPLOITATION and ALIENATED RESOURCES! #ConfirmSecGina." Twitter, February 26, 2017. Accessed January 22, 2018. https:// twitter.com/Joeltaborasj/status/836119451569274881.

Villegas, Socrates. 2016. "Prophets of Truth, Servants of Unity." CBCP News, May 1, 2016. Accessed June 16, 2021. https://cbcponline.net/prophets-of-truth-servants-of-unity/.

———. 2017. "Lord Heal Our Land." Catholic Bishops' Conference of the Philippines, September 12, 2017. Accessed January 24, 2018. http://cbcpnews.net/cbcpnews/lord -heal-our-land/.

Wilkinson, Olivia. 2017. "'Faith Can Come In, but Not Religion': Secularity and its Effects on the Disaster Response to Typhoon Haiyan." *Disasters* 42 (3): 459–74. doi: 10.1111 /disa.12258.

Williams, Sean. 2018. "How the Catholic Church Is Fighting the Drug War in the Philippines." *America Magazine*, January 25, 2018.

World Evangelical Alliance. 2015. "Global Evangelical Leaders Welcome Paris Climate Agreement." World Evangelical Alliance, December 18, 2015. Accessed January 24, 2018. http://worldea.org/news/4630/global-evangelical-leaders-welcome-paris-climate -agreement-as-historical-accomplishment.

Youngblood, Robert L. 1987. "The Corazon Aquino 'Miracle and the Philippine Churches." *Asian Survey* 27 (12):1240–55.

———. 1990. *Marcos against the Church: Economic Development and Political Repression in the Philippines*. Ithaca, NY: Cornell University Press.

DAVID T. BUCKLEY is Associate Professor of Political Science and Paul Weber Endowed Chair in Politics, Science, and Religion at the University of Louisville. His research focuses on the often-contentious relationship between religion and democracy. His book, *Faithful to Secularism: The Religious Politics of Democracy in Ireland, Senegal, and the Philippines* (Columbia University Press 2017), received the International Studies Association's 2018 Book Award for Religion and International Relations. He was a Council on Foreign Relations International Affairs Fellow (2016–2017), serving as Senior Advisor in the Department of State's Office of Religion in Global Affairs.

2

"THE EARTH IS THE LORD" OR "GOD IS A TRINI?"

The Political Theology of Climate Change, Environmental Stewardship, and Petroleum Extraction

J. Brent Crosson

THE YEAR 2017 WAS A PARTICULARLY DESTRUCTIVE ONE for the Caribbean region. The succession of powerful hurricanes that leveled Caribbean islands underscored one of the most important global issues today: climate change. As Hurricane Harvey battered Texas, the Caribbean Community (CARICOM) Secretary General warned US President Donald Trump that climate change affects "nations across the globe" (Richards 2017). Although US government officials ruled out any talk of climate change in relation to hurricanes in the aftermath of Harvey, the next month illustrated the effects that rising sea-surface temperatures could have for hurricane intensification. As two category 5 storms devastated a string of Caribbean islands, the CARICOM chairman and prime minister of Grenada did not echo the denials of the US president and his cabinet officials. "There can be no question for us in the Caribbean," he said. "Climate change is an existential threat" (Mitchell, qtd. in Richards 2017).

The Caribbean will most likely bear the greatest brunt of the effects that rising sea temperatures will have on tropical cyclones in the Atlantic Ocean. Nevertheless, in the face of such distress, at least one Caribbean island often seemed to confidently except itself from the strengthening global narrative of climate disasters: Trinidad. As in the United States, such exceptionality was attributed not just to politics (in the usual sense of the word) but to theological claims about God's sovereign power (and his national sympathies). In the wake of the devastating storms of 2017, Trinidadians could repeat an adage that has often signified their supposed immunity to hurricanes: "God is a Trini." As one writer explained this position, "Let me put forth moments of [God's] Trini-ness: Hurricane Joyce of 2000, Bret of 1993, and Arthur and Fran of 1990—which were all on a collision course with Trinidad, when at the last second they turned north with their tails between their legs. God used the same stick he used to pelt mangoes with in his childhood to turn those storms north. Only a Trini God could do that" (Warner 2010).

"God is a Trini" expresses Trinidad's alleged exceptionality in relation to the rest of the Caribbean; this southernmost island of the Antilles is mostly immune to hurricanes and gifted with oil and gas reserves that makes its gross domestic product (GDP) relatively high (and its dependence on tourism relatively low). Bucking the stereotype of the hurricane-prone beach destination, industrialized Trinidad allegedly lies just below the hurricane belt and has often claimed the divine favor of geographic positioning and geological resources. God, in this narrative, is Trinidadian.[1]

Such a narrative rang hollow for those Trinidadians with whom I have worked for more than a decade. Two tropical storms hit their region of southern Trinidad in the early part of the 2017 hurricane season. Although these storms were not nearly as strong as the hurricanes that would later affect islands farther to the north, the lack of preparation for these tropical cyclones led to significant damage. The galvanized metal roof on the house of the family I lived with was torn completely off, with parts of it left dangling from electrical poles (the roof was only held to roof beams by the weight of concrete blocks in places). The daughter of the family sent me photos via WhatsApp of their metal roofing dangling in the wind among

---

1. Throughout this work I refer to Trinidad rather than Trinidad and Tobago as the national context for these claims about God's nationality. Although forming one nation-state, Trinidad and Tobago have very different positions in relation to both hurricanes and the oil and gas economy. Tobago's main industry is tourism, and the needs of this industry often sit at odds with the priorities of oil and gas development. Tobago is also more firmly within the hurricane belt and has experienced significant levels of damage from hurricanes in fairly recent memory. Moreover, Tobagonians themselves are often wary of identifying with Trinidad for various reasons (including industrial Trinidad's reputation for crime and hubris). I thus avoid conflating the two islands throughout this chapter.

the electrical wires. Across the street, the shed in which one of my closest friends did his mechanic work was leveled, and the second story of his decrepit board home was flattened. We scrambled to raise the money for a new roof in the midst of the rainy season, and, as in the rest of the Caribbean, the damage made residents speculate on a new era of extreme climate events. Rather than claiming that God was a Trini, many of my lower-class interlocutors used an alternative conception of God derived from the African Caribbean Spiritual Baptist religion to interpret these events. This view asserted that "the Earth is the Lord" and that extraction and consumption of the Earth's resources could cause earthly disasters.

This chapter uses discourse analysis and ethnography to compare these two divergent "political theologies" of God and their respective impacts on transnational discourses of climate change.[2] Throughout this chapter, I use the term *political theology* to express close linkage between ideas about divine power and sovereign power (whether of God, the nation-state, international environmental regulations, or the Earth). My focus, therefore, is not on religion as an ideology that influences (or fails to influence) attitudes toward environmental issues—an approach that has characterized most work on the relationship between religion and climate change.[3] Rather than an ideology or set of beliefs, I am mainly interested in religion as a statement about the location of sovereign power. In this reckoning, religion and politics are not two separate ideologies that might influence each other. If both theology and politics are assertions about the proper location of sovereignty—defined, ultimately, as the power to dispose of life—then the idea that religion and politics are separate entities does not really make much sense. The statements "God is a Trini" or "the Earth is the Lord" are at once cosmological, political, and theological assertions that have important effects on the ways that anthropogenic environmental change is articulated.

---

2. On the contemporary rearticulations of political theology and the term's fraught genealogy, see De Vries and Sullivan (2006).

3. Political ideology and religious affiliation have been the two key variables in studies of attitudes on climate change. Most of these studies have focused on the United States and on "Judeo-Christian" religious affiliations. These studies have sought to explain the exceptionality of the United States and its low levels of concern about climate change in relation to other "developed countries." For a good summary of these previous approaches, which treat religion and ideology as separate variables, see Arbuckle (2017). Arbuckle's study attempts to correlate these two separate variables, finding that (at least among "Judeo-Christian" traditions in the United States) religion generally has a moderating effect on the primacy of political ideology in determining attitudes toward climate change (with Judaism being the exception to this rule).

Both of these political/theological statements diverge from dominant narratives of the relation between religion and climate change, which have been shaped not simply by a notion of religion and politics as separate ideologies but also by a US evangelical or "Judeo-Christian" perspective on these issues.[4] These narratives often assert that evangelical Christians' belief in an imminent apocalypse (Guth et al. 1995) or a "Judeo-Christian" emphasis on human dominion over nature (White 1967) lead to lack of commitment to political initiatives to mitigate climate change. In more hopeful accounts, a conservative (US) evangelical viewpoint is counterpoised to a liberal Christian political theology of "stewardship," in which God's gift of managerial dominion over nature to humans can also inspire environmental responsibility (see Hand and van Liere 1984; Eckberg and Blocker 1989; Hall 1990; Scharper 1994; Sherkat and Ellison 2007). Although many authors have touted the promise of stewardship as a theological rationale for environmental concern, other studies have shown that evangelical commitments to stewardship do not generally alter weak commitments to doing something about climate change among US citizens (see Sherkat and Ellison 2007). Both stewardship and end times eschatology assume a hierarchical relationship with God at the top, human stewards in the middle, and the Earth at the bottom. Indeed, weak commitment to climate action among evangelicals has been attributed to the perception that environmentalism is a form of "Earth worship" that inverts this hierarchy (see, e.g., Carr et al. 2012). "God is a Trini" and "the Earth is the Lord" present a very different picture of contrasting political theologies and their relation to both the Earth and climate change.

"God is a Trini" asserts that Trinidadians can avoid disasters and misfortunes and that God's gift of natural resources inspires their rightful consumption as a sign of Trinidad's divine blessing. Rather than evangelical fundamentalism, this political theology is perhaps closer to the charismatic "prosperity gospel" that has exerted enormous power in the Americas over the past few decades. Within the volatile global market for oil and gas, however, rising and falling prices create what I call the "prosperity-austerity gospel" for postcolonial petrostates like Trinidad and Tobago. The political theology of God's Trini-ness can quickly become a mark of hubris and a

---

4. Throughout this chapter, I put "Judeo-Christian" in quotation marks to foreground the constructed, politicized, and historically specific nature of this category in post–World War II Euro-American foreign policy and religious imaginings. On the construction of the category "Judeo-Christian," see Nathan and Topolski (2016).

call for an equally Protestant-inflected theology of discipline, conservation, thrift, fiscal cuts, neoliberal reforms, and "good stewardship" of the nation's resources, especially when global commodity prices fall. Rather than a sanguine narrative of stewardship, the delegation of God's sovereignty over a resource-rich creation to Trinis is both a blessing and a curse, inspiring rising GDPs and heavy dependency on global commodity prices or foreign capital.

The people I knew at my field site, however, took a different position that questioned the theological premises of both prosperity and stewardship. By asserting that "the Earth is the Lord," my interlocutors conceived of creation as something other than a possession of God loaned to human stewards (and the nation-state that is supposed to represent them). Rather than focusing on greenhouse gases and atmospheric conditions, "the Earth is the Lord" pointed toward a bodily injury to God—principally, the extraction of oil, which was conceived of as "the blood of the Earth." These two divergent political theologies highlighted the importance of two different kinds of relationships between religion and climate change: one was modeled as much on the charismatic prosperity gospel as on a "Protestant ethic" of resource conservation, and the other was modeled on what US evangelicals might call "Earth worship" but which my interlocutors saw as part of an authentic African Christianity.

"The Earth is the Lord" structures my interlocutors' concern about both climate change and petroleum extraction. Although a recent ethnography by a US anthropologist has characterized Trinis as culpable for and unconcerned about climate change and petroleum extraction (Hughes 2016), both quantitative data (a Gallup World Poll) and my ethnographic research in Trinidad over the past eleven years do not support this anthropologist's conclusion.[5] Even the idea that "God is a Trini," as I show, is as

---

5. See Lee et al. (2015) for summaries of this unprecedented 2007–2008 Gallup World Poll, which ranks Trinidad and Tobago as the country with the third highest level of concern about climate change in the world. David Hughes' (2016) study of Trinidad, *Energy Without Conscience,* argues that Trinidadians are largely unconcerned about anthropogenic climate change, with even environmental activism focused on other issues. For example, Hughes uses the example of environmental opposition to a commuter rail line in the name of protecting local farmlands as an example of misplaced concern. The commuter rail would have lessened Trinidadians' high dependence on automotive transportation, thus prioritizing the lowering of carbon emissions over protection of farms that the rail would traverse. However, struggles for the preservation of local lands and concern about climate change are not mutually exclusive for environmental activists in Trinidad. Trinidadians have also pointed out to me that by far the main source of Trinidad's high per capita carbon emissions is not transportation but the industrial processing of hydrocarbons

easily a statement about the hubris of Trinidad and the "resource curse" of hydrocarbons as an expression of blithe unconcern. It seems to me more accurate to say that my interlocutors express their marked concern about climate change and hydrocarbon extraction in ways that are partially illegible within elite Northern discourses of climate activism that privilege atmospheric carbon. In this chapter, I show how the conception of "the Earth" as a quasi-vital body structures my interlocutors' political theologies of climate concern. Although most authors have used stewardship as a potential means by which to bridge religious and climate concerns, I close the chapter by examining confluences and divergences between "the Earth is the Lord" and the growing popularity of conceptions of the Earth as a vital being in both earth science and spirituality in the global North (see, e.g., Lovelock 1990; Latour 2017; Ruse 2013). Might the notion that the Earth can become unsettled and respond to human actions be a point at which both natural science and modern spirituality partially converge? I will argue, however, that "the Earth is the Lord" presents a different response to climate change from Gaia theory or neopagan Earth worship, one that depends more on an ethics of injury, earthly difference, and pain than animistic oneness.

Most of my interlocutors described in this chapter consider themselves Christians and Africans. Although "the Earth is the Lord" and "God is a Trini" would seem to represent radically divergent propositions, both of these self-consciously Christian political theologies hinge on the divergent interpretation of a single line from Psalm 24: "the Earth is the Lord's and the fullness thereof" (Authorized [King James] Version, Ps. 24:1). The rest of this chapter highlights the importance of these two different interpretations of this biblical passage in understanding religion and climate change beyond a framework of either US Christianity or carbon-centric climate concern.

God Is a Trini: The Earth Is the Lord's

In *The Oxford Companion to Christian Thought*, the entry entitled "prosperity" begins with the following words: "There is no one Old Testament

---

for export to Europe and North America. Trinidad's carbon emissions are thus entangled with consumption in foreign markets. Certainly, Hughes is right in pointing out the complicity with climate change in Trinidad, especially amongst middle- and upper-class persons who benefitted more significantly from the energy sector. However, concern about climate change was readily apparent amongst my lower-class interlocutors, although such concern was articulated in different terms from global discourses on carbon emissions.

attitude towards prosperity. One salient strand is the view that since 'the Earth is the Lord's and the fullness thereof' (Ps. 24:1), God rewards his faithful followers with a share of his fullness" (Hastings, Mason, and Pyper 2000, 570). This strand, although contradicted at other points in the Bible, often forms part of the scriptural basis for what has been called the "prosperity gospel" in late modern Christianity (see also, Cor 9:11; 3 John 1:2). Its premise is the idea that "the Earth is the Lord's"—a possession that can be gifted to humans as a sign of their blessed faithfulness.

As this same entry, and as a number of other scholarly sources have avowed, the prosperity gospel was a peculiarly late twentieth-century US and Pentecostal creation (Hastings et al. 2000). As with many of the late-twentieth-century theological ideas marked as American and Pentecostal, the fertile ground for these ideas has not simply been "America" but rather the Americas more broadly (and other parts of the postcolonial world). These ideas were worked out not only via US and Canadian missionary trips south but by the training of Latin American and Caribbean church leaders in the US. Such back-and-forth movement between South and North inspired a kind of dialogue about stark differences in material wealth across the Americas. Through the lens of the prosperity gospel, the increasing urban poverty in many Latin American and Caribbean countries in the late twentieth century could be attributed not to US-led capitalism and neoliberalism (as it has been in so many left-leaning invectives in the region) but to southern countries' failure to practice the right kind of Christianity. The right kind (for these neo-Pentecostals) was "born again" Protestant Christianity, meaning a break from the Catholic church and/or the forces of native religions (which include both African religions and Hinduism in Trinidad). The prosperity gospel was thus linked to an equally powerful neo-Pentecostal movement: spiritual warfare theology. African religion was the bête noire of such theology—a fact on display in the invectives of US, Haitian, Brazilian, and other Caribbean Pentecostal church leaders against African-identified forces as demons that bring poverty and misfortune to their countries (see Macedo 1990; McAlister 2014; McCloud 2015; McGovern 2012; Crosson 2019). To share in God's prosperity often meant becoming born again and battling the demons of native religions.

Pentecostal-charismatic Christianity is the fastest growing religion in Trinidad and the rest of the Caribbean. The idea that "God is a Trini" draws on some of these prosperity gospel ideas, which have detached themselves from any particular church and permeated popular discourse. Nevertheless,

"God is a Trini" also remains close to the "Old Testament" strand of prosperity identified in *The Oxford Companion to Christian Thought:* the blessing of earthly resources is a sign of the Lord's covenant with a particular nation and a certain people. Just as Yahweh was the God of Israel, a late-modern postcolonial God is a Trini. The sovereignty of a political apparatus (the state) over a certain body of territory and its wealth reflects a particular postcolonial relationship with a national god.

This convergence of prosperity, God, and nation was on display in the midst of my longest period of continuous fieldwork in Trinidad (2010–2012) when oil prices were still quite high. At a press conference in March 2012, then-Prime Minister Kamala Persad-Bissessar raised two glass canisters of "black gold" above her head, announcing a "discovery" of 48 million barrels of crude oil. Really, the oil was the result of working over older offshore fields in the Gulf of Paria between Venezuela and Trinidad, and the touted "discovery" had actually happened almost a year earlier. Nevertheless, the prime minister took advantage of the press conference to claim that the oil she held above her head proved once again that "God is a Trini." This phrase graced the front pages of national newspapers, as the smiling Hindu-Christian[6] prime minister held the nation's black gold in her hands and avowed that unprecedented prosperity lay ahead of the blessed twin-island nation (although God was a Trini, he apparently would also favor Tobago to some extent).

The subsequent fall of oil prices has replaced a national discourse of prosperity with one of austerity since 2016. A new national government has utilized this discourse of lean times to authorize a string of cuts to education and public spending. Rather than a sign of a blessed national prosperity, the phrase "God is a Trini" has increasingly reflected the hubris of the "oil-rich" island. These positions of prosperity and austerity have flipped back and forth with the rise and fall of oil prices in previous eras, most particularly with the 1970s "oil boom" and the subsequent "oil bust" that

---

6. Reflecting a broader trend, Kamala Persad-Bissessar has identified as "culturally" Hindu while being baptized in a Baptist church. Her husband is Presbyterian, reflecting the widespread missionization of Canadian Presbyterians among Indo-Trinidadians. Because all denominational schools were Christian until the mid–twentieth century—with the most prestigious secondary schools remaining Christian on the island to this day—education and advancement for Hindus was often premised on embracing some form of Christianity. Persad-Bissessar embraced her multireligious nature as a multiracial appeal that could cut across the dominant ethnopolitical line between Indians and Africans in the country.

brought the country into the hands of international lending institutions, who espoused similar discourses of austerity (see Karl 1997). The prosperity gospel could easily give way to an equally Protestant gospel of conservative economics, stewardship, discipline, thrift, and saving. The budget discussions of 2017, which coincided with the workshop on climate change in Trinidad that helped to inspire this volume, were marked not by discourses of divinely sanctioned social spending but rather by a discourse of what officials called "good stewardship" of the nation's resources (and the allegedly irresponsible "stewardship" of the previous government; Hansard 2017). This turn from prosperity to austerity actually represented two sides of the same (Protestant-imprinted) coin forged in black gold.

Thus, when the hurricane season of 2017 arrived, there had already been a marked critical turn on the question of God's Trini-ness and the blessing of fossil fuels that could dovetail with broader regional alarm around climate change and strengthened storms. On September 14, in the wake of the potent storms further to the north, the *Trinidad Guardian* published a lengthy article entitled "Climate Refugees in the Caribbean," that strongly critiqued the US government's denial of climate change but also Trinidad's divinely inspired exceptionalism:

> US President Trump is such a climate change skeptic that he decided to withdraw from the Paris Agreement. . . . And even we in Trinidad, and less so perhaps in Tobago with the memories of Hurricanes Ivan and Flora, watch the devastation that Irma has wreaked upon the rest of the Caribbean with concern, but also with a touch of distance and detachment. We are steeped in sympathy and horror, but there is always that buffer that we have—that we-are-below-the-hurricane-belt notion that "God is ah Trini."
>
> There is no real understanding that, "There, but for the grace of God, goes I", as most Antiguans must feel having escaped by the skin of their teeth from the worst horrors of the storm. Or Jamaica, still reeling from PTSD flashbacks of Ivan and Gilbert.
>
> This is our worst enemy in Trinidad—the conviction that we cannot be touched hurricane-wise. But the reality is that we are not ready for a hurricane. And we know this.
>
> We know this every time it floods in downtown Port-of-Spain from the slightest bit of rain. We know this when parts of Central are knee deep in water during the rainy season. We know this when the rivers are high from the lack of foresight to dredge the river beds during the dry season. Any evacuation plan is ludicrous when a traffic commute in and out of the city capital at rush hour is several hours long and bumper to bumper. We are not ready. We know this. But God is ah Trini, oui.
>
> This is the kind of hubris that will be our downfall. (Mair 2017)

As the author notes, a strong rain in Trinidad can produce extensive flooding and mudslides. As the experience of my interlocutors just a few months earlier during tropical storm Brett illustrated, however, even the idea that Trinidad is below the tropical cyclone belt is technically inaccurate. During the time period for which written records exist, Trinidad has been directly affected by tropical storms or hurricanes in 1810, 1884, 1891, 1892, 1933, 1974, 1990, 1993, 2002, and 2017. Probably the most destructive of these storms was the 1933 hurricane, which occurred before the practice of naming tropical cyclones. This hurricane left eleven people dead and thousands homeless, covering much of southwestern Trinidad with oil that flowed from broken wells and pipelines (Advocate 1933). A recent study by the state oil company in Trinidad showed that a hurricane today could lead to a similarly massive hemorrhaging of hydrocarbons (Shyam Dyal, cited in Hughes 2016, 136). As the esteemed University of the West Indies professor Dr. John Agard showed us during the October 2017 workshop in Trinidad that helped to inspire this volume, the very idea that Trinidad is below the hurricane belt is more myth (or theology) than empirical fact.[7] Indeed, the assumption in the *Trinidad Guardian* article that Trinidadians were simply distanced from any destruction that hurricane season was a perspective that was not shared by my lower-class, rural interlocutors, who had experienced flooding and significant destruction of property as the result of two tropical storms. The position that God is a Trini was perhaps a view from a certain class position, centered in the environs of Trinidad's northern capital.

The (equally Christian-inflected) alternative that this *Trinidad Guardian* article provides for the "hubris" of "God is a Trini" is a well-known reminder of humility: "There, but for the grace of God, go I." This adage, typically spoken after hearing of someone else's misfortune, reminds listeners that such misfortune could easily befall them. I never heard anyone use this phrase at my field site, but the sentiment of this adage was expressed in a different way—it was normal to append any statement of future plans with the phrase "[if it] please God." This more austere idea

---

7. As part of an initiative funded by the Luce Foundation on "Religion and Climate Change in Cross-Regional Perspective," American University's Center for Latin American and Latinx Studies organized and cosponsored a two-day workshop in Trinidad focused on "Small Island Vulnerabilities in the Pacific and Caribbean," hosted by the Institute of International Relations at the University of the West Indies, St. Augustine.

of God, as a figure whose ability to cut short a life or cause catastrophe could manifest at any moment, has also been used by politicians to call for a hedonistic society to seek repentance and wean itself from dependence on oil and gas revenues. In the wake of the most devastating hurricanes, for example, a member of parliament (MP) for a largely rural district of eastern Trinidad received a message/vision from God along those lines. God allegedly told the MP, Christine Newallo-Hosein, that the nation was heading for a divinely inflicted catastrophe unless its prime minister begged for forgiveness in front of Him. She promptly typed up God's words as she had heard them and delivered them to the prime minister during a ground-breaking ceremony for a controversial highway that was to pass through the district she represented. Her constituents and Trinidadian environmentalists were opposed to the highway and allegedly saw the extension as a strategy to raise the property values of government officials who owned land in the region. Newallo-Hosein was going to decline the prime minister's invitation to the sod-turning ceremony, but God told her to go and deliver His message. After handing a paper copy to the prime minister before he turned the sod, she promptly left the event.

The God of this message is angry and potentially vengeful if humans do not express repentance for their environmental destruction. Although this divine message could easily be interpreted as nothing more than a political stunt, it represents a telling conjunction of events that are familiar within Trinidad's environmental movement. The most recent focal point for environmental activism in Trinidad was another massive highway extension that was said to represent both the unnecessary destruction of the earth and the previous government's corruption. The extension was controversial both because it would drain environmentally sensitive wetlands and because it was popularly seen as another instance of what one hunger-striking environmental activist (Wayne Kublalsingh) called the "contractocracy." This contractocracy allegedly funneled government oil and gas rents to private contractors who charged overpriced amounts for shoddy infrastructural work and, in turn, financially supported political campaigns.

In the wake of the 2017 hurricane season and an intensified public discussion about climate change, Newallo-Hosein's words went beyond a critique of political corruption. This corruption, reflected in environmentally insensitive development regimes of infrastructure and energy, would inspire divine wrath. God's words of warning, she said, were not simply directed at the prime minister but also at a country on the verge of divinely inspired environmental disaster:

> Newallo-Hosein said the words spoken to the Prime Minister were for all citizens. She is now hoping that [Prime Minister] Rowley heeds the word of God and leads the nation in repentance: "My heart's desire is that the Prime Minister would heed His word, not mine, and that he would lead the nation in repentance. That he would call the churches and they would repent and call a genuine call to repentance."
>
> Newallo-Hosein indicated that citizens should take a serious look at what was happening in neighbouring islands [struck by hurricanes] and start praying.
>
> "There is a sense that there is pending danger that is coming to our nation and we keep saying God is a Trini and his mercy has been with us. We seeing all these islands [hit by hurricanes] literally stripped and this is not a time for people to say this is politics. It is a time for people to get on their knees and repent and say Lord have mercy on us," she said. (Kissoon 2017)

Once again, the hubris of "God is a Trini," which popularly expresses Trinidad's alleged immunity to natural disasters and its benefit from hydrocarbon wealth, is contrasted with the sense of impending disaster. The popular idea that God's wrath is behind natural disasters in the Caribbean has had some disturbing implications that do not necessarily inspire substantive change in regnant economic or environmental policies. US and Haitian Pentecostal and charismatic Christians alleged that the immensely devastating 2010 earthquake in Haiti was divine punishment for the island's practice of Voodoo. US televangelist Pat Robertson, echoing sentiments expressed by Haitian evangelicals themselves (Louis 2015), proclaimed that the "Voodoo ceremony" inaugurating the Haitian revolution (Bois Caïman) was "a pact with the devil" that had inspired God's wrath. The poverty of Haiti and the recent earthquake were supposed evidence of this divine vengeance for an African religious rite. I once heard the same explanation in Trinidad, from a taxi driver, for Haiti's earthquake. The austerity gospel of a vengeful God risked erasing political and economic reasons for environmental and social destruction and proposed that the solution was to repent and practice the right kind of Christianity. Prosperity and austerity hinged on God's total sovereignty and the idea that the forces of the Earth—both resources and disasters—were signs of either divine favor or divine anger.

Like the stewardship model, the prosperity-austerity gospel leaves the hierarchical Great Chain of Being—with God at the top, human stewards in the middle, and the Earth at the bottom—very much intact.[8] This is precisely why authors have proposed stewardship as an environmental ethos that might be palatable to Christian climate skeptics (Carr et al. 2012). Other studies, however, have shown no correlation between an embrace of Christian stewardship ethics and serious concern about climate change. The austerity gospel is a conservative political theology of conservation that sees humans as arrogant, sinful actors who must repent to God. The material outcomes of this repentance are unclear. Austerity could simply authorize the fiscal belt-tightening (i.e., slashing of social services) that hurts lower-class Trinidadians most. It could just as easily lead to moral condemnation and persecution of African-identified religions. If the Earth is the Lord's—a possession that God delegates or withholds in line with His favor—then what matters more is one's relationship with a transcendent God rather than with the Earth itself.

The Earth Is the Lord

In the Trinidad English Creole spoken by my interlocutors, possessive nouns are not usually formed by adding an audible *s*. Instead of saying, "The Earth is the Lord's," one would technically say in spoken English Creole, "The Earth is the Lord own." The popular biblical maxim that "the Earth is the Lord's" becomes "the Earth is the Lord" in spoken English at my field site. For the autochthonous Afro-Caribbean Christian religion known as Spiritual or Shouter Baptism, "the Earth is the Lord" is an extremely popular maxim that weaves in and out of everyday conversations beyond the walls of a particular church or denomination. The ambivalence of possession in this maxim reflects the view of many of my interlocutors—that the material Earth is literally the Lord, a part of God.

---

8. The Great Chain of Being allegedly derives from classical conceptions of a hierarchy leading from inanimate matter (low) to ethereal spiritual beings (high). This chain was a cornerstone of medieval Christian theology and the Aristotelian natural philosophy, positing a hierarchy of dense matter and ethereal spirit. The Earth represented the densest and basest kind of matter, with the center of the Earth (and the universe) being Hell—where the densest forms of matter congregated. The planets and the sun represented more refined celestial realms that orbited around the Earth, with the outermost limits of these orbits reserved for angels and, finally, God. The Great Chain both connected these realms and arranged them in hierarchical order. This hierarchy of matter from dense (low) to ethereal (high) also organized medicine and astrology, with the humors and the planets associated with different kinds of matter. In this chapter, I use the Great Chain to reference the hierarchy of God, humans, and Earth that still obtains in much Euro-American Christian theology, even though most of us no longer suppose we live in a geocentric universe. For more on these issues, see Berry (2010) and Danielson (2009).

This view led to a corporealization of oil extraction and environmental threats, which made empathic metaphors between the health of a human body and the Earth. One of my friends—a twenty-four-year-old man of African descent named Curtis—thought that extracting oil was like extracting blood from a human body. "The more they suck oil and gas," he told me as we *limed* (hung out) across the street from the village playing field, "the weaker the Earth get." One of the only sources of formal employment in this region was manual labor on occasional seismic surveys that explored for oil and gas (fishing and farming were far more crucial and regular forms of livelihood). He had worked as a manual laborer on the drilling crew of the region's last seismic survey, which was a fairly good position—much better than bushwhacking paths for seismic cable in dense forests that held venomous snakes and pipe-gun traps protecting marijuana fields. The drilling crew's labor began after these paths were cleared, as the land surveys used dynamite charges buried in holes to create shockwaves powerful enough to penetrate the Earth and echo back to the cables' geophones. Curtis told me how they had to drill dynamite holes into a group of mountains known as the "Three Sisters." These mountains are the largest in the region, and their tripartite grandeur had inspired Columbus to christen the island after the Holy Trinity—*la Trinidad*. Curtis said he would never do such work again. Drilling into those mountains, he said, "ain't right at all. I is a man does deal with nature. What is greater than man?" he asked me rhetorically. "Them mountains," he said, answering his own questions. "[They have been] there long time, since the Earth begin. The Earth is the Lord."

When Curtis defined himself as a man who dealt with nature, he meant that he tried to pay attention to scales of time and force that were "natural" rather than human. God represented these scales that far exceeded the span of a human life—the time of mountains, rocks, and other geological forces. A somewhat similar view was expressed to me a few months earlier, just as the seismic crew was about to start drilling into the Three Sisters. A farmer and spiritual worker named Roland, who was in his early sixties and lived at the foot of the Three Sisters, claimed that by drilling into the mountains, the seismic crew would provoke the next apocalypse. Roland showed me a map of the seismic survey that he had acquired from someone who worked on the crew. He spread it out on the hood of a broken-down car in front of his house. Roland pointed out the Three Sisters and began to explain why Noah's ark had been constructed on those mountains. The Bible said the ark was built of green (i.e., freshly cut, untreated) wood, and the only green

Figure 2.1. Roland, 2011. Photograph by Gigi Gatewood.

wood that could float, he told me, was the wood of the silk cotton tree (*Ceiba pentandra*), which was native to the American tropics and was sacred in African Trinidadian religious practices.[9] He said the next apocalypse would also center around the Three Sisters; the insult of humans drilling into their bodies marked the beginning of the end times. He was trying to get a group of residents together to block the dirt road that the seismic crew would have to use to access the Three Sisters.

Echoing another corporeal metaphor that Curtis had used to criticize oil extraction, Papoy, an Afro-Trinidadian farmer in his late fifties, compared oil companies to *soucouyant* (vampires). We were seated on the fallen log that served as an impromptu hangout spot and bench behind a local corner store. Papoy pointed to his toe and explained that the oil companies, or soucouyants, could be sucking blood out of his head from an extraction point down there in his foot. With this corporeal metaphor of parasitic extraction, he was explaining how the oil that lay underneath their house and land was being extracted by platforms off the coast, bypassing any remunerative interaction with local residents. Indeed, as Curtis had also avowed, energy companies had made the decision to extract onshore oil via offshore platforms to circumvent a local population that was seen as recalcitrant. The most recent seismic survey had been marked by demonstrations and road blockages over labor issues, and the local liaison for the major energy firm operating in the region told me (off the record) that Rio Moro had a reputation in the industry for labor resistance. Pointing from his toe to the earth below him, Papoy told us that he could see his crops growing stunted as they sucked the oil. The oil was the blood of the Earth, he said, and the more they sucked, the weaker the body of the Earth was. This interpretation meant the soil was less vital and less able to support crops. Echoing a theory I heard from other farmers, he speculated that the fertilizer that was then necessary to grow crops led to vegetables that were less nutritionally rich. This was so because nitrogen fertilizer was made out of the natural gas (drawn from beneath their feet) that energy-intensive plants in western

---

9. Not being overly familiar with the Bible, I simply assumed that God had instructed Noah to use green wood in the King James Version. Later, I found that the instructions in the King James Version that my interlocutors use are to make the ark out of "gopher wood," from the original Hebrew word *gofer*—a word otherwise unknown in the Hebrew language. Not finding a proper translation, the King James Version simply left the word more or less untranslated. Roland then read "gopher," an unknown word to him, as "green" and elaborated a flood story centered in his own sacred geography of the Three Sisters and silk cotton trees.

Figure 2.2. Papoy in his garden, 2011. Photograph by Olivia Fern.

Trinidad processed into ammonia. With a higher "gas" content, vegetables were less nutritive and provoked excessive flatulence. Although this theory might seem strange, it was a common one, and its analysis of the making of modern nitrogen fertilizers was quite accurate. Papoy asked the others if they had noticed how much they farted when they "eat chemical"— meaning the consumption of store-bought vegetables and foods grown with chemical fertilizers. There were murmurs of agreement.

These theories about injury to earthly and human bodies, flatulence, and apocalypse might seem inchoate to outsiders; however, they all reflected a root metaphor and logic—that the Earth is a vital body. David Hughes (2016) has argued that Trinidadians—even those who call themselves environmentalists—are largely unconcerned about oil and gas extraction (see note 5). Living in the world's oldest petro-state, they see oil extraction as a natural part of the landscape. This "petro-pastoral sensibility," as Hughes calls it, and a broader reliance on oil and gas industries mean that oil and gas extraction is not subject to criticism. My field site had been a site of oil extraction since the early twentieth century, but this alleged petro-pastoral sensibility was largely absent. Oil and gas extraction was far from innocuous. For one thing, it threatened the two dominant livelihoods in the region—fishing and farming. Fisheries were threatened by both regular oil spills and the submarine explosions that were used in offshore seismic surveys, which have been shown to significantly affect fish populations (despite contested industry-funded studies; see, e.g., Surtees 2013; Fitzgibbon et al. 2017; Paxton et al. 2017). By sucking the blood out of the Earth, fossil fuel extraction also affected the vitality of soils, according to local food-crop farmers like Papoy. For those who cultivated marijuana instead of food crops, the seismic surveys that criss-crossed the region's dense forests had the potential to disrupt their activities (although the forest knowledge of local manual laborers on seismic crews helped avoid this). Perhaps Hughes's conclusions hold true for Trinidadians who have had more to gain from the energy sector and state development projects. My interlocutors tended to see both the state and the oil companies as parasitic entities that actually threatened their livelihoods, particularly food-crop farming, fishing, and the illicit cultivation of marijuana.

At a more theological level, the conception of the Earth as the Lord meant that drilling, dynamiting, and extracting from the Earth was a bodily injury or an insult to God that had repercussions. I conclude this chapter by elaborating on my interlocutors' political theology of "the Earth is the Lord" as an alternative to pervasive eco-theologies of stewardship and Gaia.

From Anthropocene to Injury?

Following some well-worn racial stereotypes, it would be easy enough to represent Afro-Christian Spiritual Baptist views as animist survivals of African tradition that predated Christianity and/or modernity. Rather than relegating "the Earth is the Lord" to a Romantic past of animism or premodern Earth worship, I want to close by examining the resonance of this view with a recent theory in earth sciences and contemporary movements of spirituality in the global North. Both science and spirituality have increasingly converged around an enigmatic earth goddess resurrected from Greek mythology—Gaia. This resurrection owes much to "the Gaia hypothesis," proposed by chemist James Lovelock in the late 1960s and later developed with the influential evolutionary theorist and microbiologist Lynn Margulis (among many other collaborators). This hypothesis made the seemingly novel proposition that the Earth is a complex and self-regulating life form. Originally ridiculed by the scientific community, the idea of a living Earth moved from hypothesis to respected theory over the past three decades, as it formed the basis for empirical predictions and experimental testing.

Religion has played a key role in the reception of the Gaia hypothesis (see Ruse 2013). One of the major criticisms of "Gaia theory," leveled by Stephen Jay Gould and Richard Dawkins (among others), is that it reflected an untenable mixing of religion and science, remaining a fanciful metaphor rather than a scientific theory (Ruse 2013). Specifically, critics have seen the Gaia theory as imbued with neopagan or New Age deifications of the Earth. Indeed, such spiritual groups have also taken up the ancient Greek goddess as a central object of veneration and were some of the earliest enthusiasts of the Gaia hypothesis (Ruse 2013). Strangely enough, these avowedly secular criticisms of Gaia theory, as a thinly veiled form of neopaganism, echo evangelical Christian condemnations of environmentalism as a form of pagan Earth worship (Carr et al. 2012). Africana and indigenous religions (or Euro-American representations of these religions) are often taken as exemplars of this Earth worship for a wide range of Euro-American actors, including New Agers and neopagans in the global North, romantic environmentalists, and evangelical or scientific critics of the Earth's alleged deification.

Although many models of ecological stewardship, anthropogenic climate change, or a new geological age of the Anthropocene place humans front and center as the key agents in environmental change, Gaia theory has often depicted humans as a relatively unimportant life form (Lovelock 1979). Margulis and Sagan (1986) suggest that humans may simply play the role of incubators for the microorganisms that are important for the Gaia organism. This has led some Gaia theory proponents to suggest that human pollution or extinction is relatively unimportant for the living Earth (see Scharper 1994). For this reason, the relationship between Gaia theorists and environmentalism has often been both rocky and romantic, with Havelock strongly supporting fracking and calling more dire predictions about climate change or human environmental destruction "alarmist" or "religious" rather than fact based (Vaughan 2014).

This demoted place of human agents in Gaia theory has led ecologically minded Christian theologians to insist that stewardship is an ethically superior framework (see Scharper 1994). Stewardship keeps humans front and center as actors, calling them to account for climate change and insisting that they have a responsibility to do something about it. Nevertheless, the practical outcomes of a commitment to Christian environmental stewardship are uneven, and quantitative studies have shown that the commitment to environmental stewardship has no impact on serious concern about climate change among US Christians (Sherkat and Ellison 2007). Perhaps, the reason for this lack of impact can be found in the word *steward* itself. Because the steward is middle management in the Great Chain of Being, the word arguably functions much as stewards did in the past. The word *steward* originally denoted an overseer of workers and a manager of servants or property in modern English.[10] In stewardship, God assumes the place of the property owner, and the Earth assumes the position of the servants, slaves, land, natural resources, or objects whose care is delegated to the steward.

Following this (admittedly uncharitable) analogy, stewardship can sound like paternalistic apologies for slavery, which centered on Whites' alleged role as stewards or caretakers of slaves. Rather than changing the theological, racial, and moral Great Chain of Being that supported the

---

10. "Steward, *n*. (1)," *Oxford English Dictionary*, accessed March 21, 2021, https://www-oed-com .ezproxy.lib.utexas.edu/view/Entry/190087?rskey=CP9zlA&result=1#eid.

institution of slavery, such apologies relegated slaves to an inert, defenseless position. Some authors (Mouhot 2011; Hughes 2016) have critically invoked this Great Chain to describe the contemporary predicament of energy-intensive environmental destruction, comparing human exploitation of hydrocarbons with the exploitation of enslaved human energy on sugar plantations. Although this comparison has some serious limitations—for one thing, it makes an extremely problematic analogy between the exploit-ation of oil and slaves—it seems important to note that both of these master narratives of capitalist energy exploitation had their start in the Caribbean (see Mintz 1985; Williams 1944/1996; Trouillot 2002). In Trinidad, sugar plantations and the world's earliest oil industry coincided (Hughes 2016). Perhaps it is worth listening to the autochthonous political theologies that emerged at this crossroads of modern energy and racial systems.

Spiritual Baptist attitudes toward the Earth provide such an autoch-thonous Afro-Caribbean political theology that disrupts the hierarchical relation of stewardship and the Great Chain of Being. "The Earth is the Lord['s]" also provides a different orientation from Gaia theory—one that is based more in an ethics of bodily injury than the notion of a self-regulating, resilient superorganism that includes humans. Precisely because the Earth is a being that does not include humans (because it is divine and nonhu-man), it can experience anthropogenic injury (see Latour 2017). Ethically speaking, empathy's root is the ability to experience the pain of an other, and thus empathy is rooted in some notion of both separable personhood and relationality. This ethics of injury forms the basis not only for empathy but for new legal regimes that, despite many challenges in implementation, define the Earth as a person with rights (see Daly 2012). Ranking just above Trinidad and Tobago in terms of serious concern about climate change (see Lee et al. 2015), Ecuador has begun to explore the legal consequences of making the Earth into a nonhuman juridical person.

Empathy (and injury) involves difference as much as sameness—the existence of at least two separable persons. The extraction and burning of hydrocarbons surely has caused bodily injury to a great number of humans; for many others, filling up their gas tanks, drinking from an aluminum can, or running air conditioning are not physically painful experiences (in fact, these actions often afford embodied comforts). Empathy requires not simply experiencing physical pain in one's own body but making an ethical leap to feel the pain in an other's body. To the degree that one comes close to this pain, paternalistic sympathy can become partially shared empathy.

The sharing is always partial because rather than forming one superorganism (i.e., Gaia), the partners of empathy are, by definition, different beings. It is important to critique and blur a divide between nature and culture or between human and nonhuman (see, e.g., Haraway 2003; Latour 2017); however, minimizing the divisions between different forms of life (and differently raced or gendered human actors) also diffuses responsibility and the locus of injury (see also Caluya 2014; Malm and Hornberg 2014). Banal environmentalist/neopagan representations of Gaia as a pregnant woman with a globe for a fetus or overgeneralized conceptions of the Anthropocene as a geological age of "mankind" risk erasing the differences that matter for an ethics of bodily injury. Even more sensitive accounts of Gaia theory might run this risk. In his advocacy of Gaia theory as the basis for a response to global warming, Latour (2017) builds on his earlier work to argue that this theory dissolves the modern dichotomy between human culture and nature, enfolding nations, deserts, and oceans in a global assemblage. In contrast, I argue that the nullification of a distinction between nature and humans (or between different human groups or nations) might preclude forms of response-ability based on the recognition of injury to an other. Erasing Western man's othering of nature (or humans) precludes the differences that are the basis for injury as well as responsibility and sovereignty. If the Earth is the Lord, then it is something other than human.

By drawing parallels between disease in human bodies and energy extraction from the Earth, my interlocutors did not think that they were a part of the Earth but elaborated ethical positions based on empathy across difference and the possibility of nonhuman personhood and sovereignty. While some of their metaphors were certainly anthropomorphic, they were not anthropocentric. Anthropomorphism allowed an analogy across sensate bodies that could turn oil into blood (or natural gas into flatulence) without turning the Earth into a human (or a ward of mankind). Because my interlocutors were low personal energy consumers whose livelihoods were often directly threatened by hydrocarbon extraction, it might have been easier for them to draw analogies between their injuries and those that the Earth experienced as a result of extraction. Nevertheless, such ethics are by no means an impossible basis for broader attempts to address climate change, which must ultimately assign responsibility for injury in political, economic, moral, and theological terms.

For now, at least, it is not theology that seems to count but proximity to pain. White US evangelicals living at the center of a global empire in northern climates (or in US coastal areas that have good access to disaster relief) are farther from the immediate pain that climate change will help inflict. In contrast, the Caribbean is at the center of this pain. Levels of education, development, and religiosity have little broad correlation with concern about climate change, according to a global survey, but the perception of risk does (Lee et al. 2015). In the wake of the 2017 hurricane season, the Caribbean—a profoundly Christian and religiously conservative region— often had no problem accepting that climate change is a crucial concern. For others, the empathic leap is greater. To make this leap, it might help to accept that energy, in one way or another, is injury.

A political theology of "the Earth is the Lord" also implies a different conception of both divine and political sovereignty. The sovereignty of God in the US Christian theologies that have dominated discussions of religion and climate change seems to assume that God is above environmental injury. Perhaps this conception of sovereignty is related both to a conception of God as above nature and a notion of the United States's "monopoly on the legitimate use of force" (to use Max Weber's classic definition of sovereignty) as a global right. That this right has often been conceived of as the militaristic spread of "freedom" as "God's gift to the world" speaks to the entanglement of political discourse and US Christian theology (see Bush 2004, 2005; Saiya 2012).

Nevertheless, it would be difficult indeed for a Trinidadian (or the resident of any contemporary Caribbean nation-state) to assume that the military of their nation exerted such sovereignty across much of the globe. The focus of US military sovereignty, at least at the level of public discourse, is not on the injuries that the exercise of such sovereignty implies but on the freedom or protection that such militarism confers. The ostensibly injury-free sovereignty of God and nation might, seem self-evident from a US Christian perspective. For my interlocutors in southern Trinidad, divine and political sovereignty that includes the real possibility of significant injury often made more sense. Such a view could draw on the theologies of West African religions, in which more-than-human powers require the care and reciprocity of humans (see, e.g., Pérez 2016). Such a view might also point toward the performative or "hybrid" nature of sovereignty in the Caribbean, which defies the implicitly monotheistic and statist conceptions of Western political theorists (Jaffe 2013; Kivland 2012; Singh 2012). To some extent, however, most religions require some form of reciprocity between humans and more-than-human power, and all sovereignties are performed by more actors than those typically recognized as the state. The ways in which such reciprocity is recognized between humans and God(s) or people and the state depend on contexts of power. Reasons for concern about climate change might be as geopolitical as theological, with these concerns implying different conceptions of sovereignty than those encoded in US Christian theology or Western political theory.

Discussions of climate change and religion have been dominated by a US perspective on both theology and politics, but I have suggested that we need to intensively explore and compare other contexts. In this chapter, I have also argued that we need to begin treating religion and political ideology not as separate variables but as intertwined conceptions of the location and character of sovereignty (whether the sovereignty of God[s], nations, nature, presidents, or international institutions). Such an approach provides one path forward for analyzing pressing issues of religion, climate change, and politics.

References

Advocate. 1933. "Hurricane in Trinidad." *Advocate*, July 3, 1933. Accessed February 10, 2018. https://trove.nla.gov.au/newspaper/article/68029063.

Arbuckle, Matthew. 2017. "The Interaction of Religion, Political Ideology, and Concern about Climate Change in the United States." *Society and Natural Resources* 30 (2): 177–94. doi:10.1080/08941920.2016.1209267.

Beliso de Jesús, Aisha. 2015. *Electric Santería*. New York: Columbia University Press.

Berry, Evan. 2010. "Great Chain of Being." In *Encyclopedia of American Environmental History*, edited by Kathleen Bronsan. New York: Facts on File.

Bush, George W. 2004. Address before a Joint Session of the Congress on the State of the Union, January 20.

———. 2005. Second Inaugural Address, January 20.

Caluya, Gilbert. 2014. "Fragments for a Postcolonial Critique of the Anthropocene: Invasion Biology and Environmental Security." In *Rethinking Invasion Ecologies: Natures, Cultures and Societies in the Age of the Anthropocene*, edited by Jodi Frawley and Iain McCalman, 31–44. New York: Routledge.

Carr, Wylie, Michael Patterson, Laurie Yung, and Daniel Spencer. 2012. "The Faithful Skeptics: Evangelical Religious Beliefs and Perceptions of Climate Change." *Journal for the Study of Religion Nature and Culture* 6 (3): 276–99.

Crosson, J. Brent. 2019. "Catching Power: Problems with Possession, Sovereignty, and African Religions in Trinidad." *Ethnos*. 84 (4): 588–614. doi:10.1080/00141844.2017.1401704.

Daly, Erin. 2012. "The Ecuadorian Exemplar: The First Ever Vindications of the Constitutional Rights of Nature." RECIEL 21 (1): 63–66.

Danielson, Dennis. 2009. "Myth 6: That Copernicanism Demoted Humans from the Center of the Universe." In *Galileo Goes to Jail and Other Myths about Science and Religion*, edited by Ronald Numbers, 50–58. Cambridge, MA: Harvard University Press.

De Vries, Hent, and Lawrence E. Sullivan, eds. 2006. *Political Theologies: Public Religions in a Post-secular World.* New York: Fordham University Press.

Eckberg, Douglas Lee, and T. Jean Blocker. 1989. "Varieties of Religious Involvement and Environmental Concerns: Testing the Lynn White Thesis." *Journal for the Scientific Study of Religion* 28:509–17.

Fitzgibbon, Quinn, Ryan Day, Robert McCauley, Cedric Simon, and Jayson Semmens. 2017. "The impact of seismic air gun exposure on the haemolymph physiology and nutritional condition of spiny lobster, *Jasus edwardsii.*" *Marine Pollution Bulletin* 125 (1–2): 146–56.

Guth, James L., John C, Green, Lyman Kellstedt, and Corwin E. Smidt. 1995. "Faith and the Environment: Religious Beliefs and Attitudes on Environmental Policy." *American Journal of Political Science* 39 (2): 364.

Hall, Douglas J. 1990. *The Steward: A Biblical Symbol Come of Age.* Eugene, OR: Wipf & Stock.

Hand, Carl M., and Kent D. van Liere. 1984. "Religion, Mastery-over-Nature, and Environmental Concern." *Social Forces* 63:555–70.

Haraway, Donna. 2003. *The Companion Species Manifesto.* Chicago: Prickly Paradigm.

Hastings, Adrian, Alistair Mason, and Hugh Pyper, eds. 2000. *The Oxford Companion to Christian Thought.* New York: Oxford University Press.

Hughes, David. 2016. *Energy without Conscience: Oil, Climate Change, and Complicity.* Durham, NC: Duke University Press.

Jaffe, Rivke. 2013. "The Hybrid State: Crime and Citizenship in Urban Jamaica." *American Ethnologist* 40 (4): 734–48.

Karl, Terry Lynn. 1997. *The Paradox of Plenty: Oil Booms and Petro-States.* Berkeley: University of California Press.

Kissoon, Carolyn. 2017. "God Came to Me, with Message for Rowley." *Trinidad Express,* September 27, 2017.

Kivland, Chelsea. 2012. "Unmaking the State in Occupied Haiti." *Political and Legal Anthropology Review* 35 (2): 248–70.

Latour, Bruno. 2017. *Facing Gaia: Eight Lectures on the New Climatic Regime.* Cambridge, UK: Polity.

Lee, Tien Ming, Ezra Markowitz, Peter Howe, Chia-Ying Ko, and Anthony Leirserowitz. 2015. "Predictors of Public Climate Change Awareness and Risk Perception around the World." *Nature Climate Change* 5:1014–20. https://www.nature.com/articles /nclimate2728.

Louis, Bertin. 2015. *My Soul Is in Haiti: Protestantism in the Haitian Diaspora of the Bahamas.* New York: New York University Press.

Lovelock, James E. 1979. *Gaia: A New Look at Life on Earth.* New York: Oxford University Press.

———. 1990. "Hands Up for the Gaia Hypothesis." *Nature.* 344 (6262): 100–2.

Macedo, Eder. 1990. *Orixás, Caboclos e Guias: Deuses ou Demônios?* Rio de Janeiro: Universal Produções.

Mair, Caroline. 2017. "Climate Refugees in the Caribbean?" *Trinidad Guardian,* September 14, 2017. Accessed February 10, 2017. http://www.guardian.co.tt/life-lead/2017-09-14 /climate-refugees-caribbean.

Malm, Andreas, and Alf Hornborg. 2014. "The Geology of Mankind? A Critique of the Anthropocene Narrative." *Anthropocene Review* 1 (1): 62–69.

Margulis, Lynn, and Dorion Sagan. 1986. *Microcosmos: Four Billion Years of Evolution from Our Microbial Ancestors*. New York: Summit.

McAlister, Elizabeth. 2014. "Possessing the Land for Jesus." In *Spirited Things: The Work of "Possession," in Afro-Atlantic Religions*, edited by Paul C. Johnson, 177–206. Chicago: University of Chicago Press.

McCloud, Sean. 2015. *American Possessions: Fighting Demons in the Contemporary United States*. New York: Oxford University Press.

McGovern, Mike. 2012. "Turning the Clock Back or Breaking with the Past? Charismatic Temporality and Elite Politics in Côte d'Ivoire and the United States." *Cultural Anthropology* 27 (2): 239–60.

Mintz, Sidney. 1985. *Sweetness and Power: The Place of Sugar in Modern History*. New York: Penguin.

Mouhot, Jean-François. 2011. "Past Connections and Present Similarities in Slave Ownership and Fossil Fuel Usage." *Climate Change* 105:329–55.

Nathan, Emmanuel, and Anya Topolski. 2016. "The Myth of a Judeo-Christian Tradition: Introducing a European Perspective." In *Is There a Judeo-Christian Tradition? A European Perspective*, edited by Nathan Emmanuel and Topolski Anya, 1–14. Berlin; Boston: De Gruyter.

Parliament of Trinidad and Tobago. 2017. "Appropriation (Financial Year 2018): Unrevised." October 12, 2017. Accessed February 10, 2018. http://www.ttparliament.org/hansards/hh20171012.pdf.

Paxton, Avery, Christopher Douglas, P. Nowacek, Julian Dale, Elijah Cole, Christine Voss, and Charles Peterson. 2017. "Seismic Survey Noise Disrupted Fish Use of a Temperate Reef." *Marine Policy* 78 (April): 68–73.

Pérez, Elizabeth. 2016. "Religion in the Kitchen: Cooking, Talking, and the Making of Black Atlantic Traditions." New York: New York University Press.

Richards, Peter. 2017. "Caribbean Wobbles under Impact of Climate Change." *Trinidad Guardian*, December 30, 2017. Accessed February 10, 2018. http://www.guardian.co.tt/news/2017-12-30/c%E2%80%99bean-wobbles-under-impact-of-climate-change.

Ruse, Michael. 2013. *The Gaia Hypothesis: Science on a Pagan Planet*. Chicago: University of Chicago Press.

Saiya, Nilay. 2012. "Onward Christian Soldiers: American Dispensationalists, George W. Bush and the Middle East." *Holy Land Studies* 11 (2): 175–204. doi:10.3366/hls.2012.0044.

Scharper, Stephen. 1994. "The Gaia Hypothesis: Implications for a Christian Political Theology of the Environment." *Cross Currents* 44 (2). http://www.crosscurrents.org/Gaia.htm.

Sherkat, Darren E., and Christopher G. Ellison. 2007. "Structuring the Religion-Environment Connection: Identifying Religious Influences on Environmental Concern and Activism." *Journal for the Scientific Study of Religion* 46:87–100.

Singh, Bhrigupati. 2012. "The Headless Horseman of Central India: Sovereignty at Varying Thresholds of Life." *Cultural Anthropology* 27 (2): 383–407.

Surtees, Joshua. 2013. "Fisherfolk on Seismic Surveys: Ministry, Petrotrin Hoodwinking Country." *Trinidad Guardian*, October 29. http://www.guardian.co.tt/news/2013-10-28/fisherfolk-seismic-surveys-ministry-petrotrin-hoodwinking-country.

Trouillot, Michel-Rolph. 2002. "North Atlantic Universals: Analytical Fictions 1492–1945." *South Atlantic Quarterly* 101 (4): 839–58.

Vaughan, Adam. 2014. "James Lovelock: Environmentalism Has Become a Religion." *Guardian*, March 30, 2014. https://www.theguardian.com/environment/2014/mar/30/james-lovelock-environmentalism-religion.

Warner, Ayinde. 2010. "Why God Is Not a Trini." *Outlish Magazine*, November 8, 2010. Accessed February 10, 2018. http://www.outlish.com/why-god-is-not-a-trini/.

White, Lynn, Jr. 1967. "The Historical Roots of Our Ecological Crisis." *Science* 155:1203–7.

Williams, Eric. (1944) 1996. *Capitalism and Slavery*. Chapel Hill: University of North Carolina Press.

J. BRENT CROSSON is Assistant Professor at University of Texas–Austin. His book *Experiments with Power: Obeah and the Remaking of Religion in Trinidad* is published by the University of Chicago Press (2020). His work on race, religion, and the criminalization of African-identified spiritual work appears in journals including *Method and Theory in the Study of Religion*, the *Journal of Africana Religions*, *Ethnos*, and *Anthropological Quarterly*. Work from his next research project on climate change, African religion, and geology appears in *Cosmologics* and the edited volume *Mediality on Trial*.

3

CONTEMPORARY POLITICS OF CLIMATE CHANGE IN INDIA

Nationalism, Religion, and Science in an Uneasy Alliance

Neeraj Vedwan

Mr. Kumar is a fifty-five-year-old realtor who has lived in the Rohini neighborhood of northwest Delhi for more than two decades. He has sold many apartments in a residential complex in which he also resides. According to him, among the factors that draw people to the complex are affordability and spaciousness. When asked to elaborate on the latter (Naren Kumar, interview, June 23, 2018), he said, "The complex is surrounded by parks and a greenbelt (of trees and other vegetation). Most similarly priced apartment complexes are congested and lack open space." Among other attractive features, he describes the availability of drinking water and electricity, both of which can be erratic in many places in Delhi. In further conversation, Mr. Kumar said that he expects water scarcity and pollution to worsen in the future. He attributes this to the increasing population and extravagant lifestyles of the middle and upper classes and to climate change. When asked about the latter, he describes hotter summers, destructive dust storms, and winters that are increasingly mild as signs of a shifting climate. But individuals are not helpless, he said. He mentioned how he uses a reverse-osmosis–based water filter at home and is thinking of buying an air purifier. Regarding what, if anything, the country should do about climate change, he took on a philosophical tone and said that after all we live in *kaliyuga* and then went on to link the "deteriorating" climate to a general decline in society, values, and morals.[1]

This chapter is devoted to exploring the linkages between policies and discourses related to climate change and the broader social-political milieu in India. Instead of representing a coherent and monolithic body of ideas, climate policies and practices are fractured and marked by multiple disjunctions. In this chapter, I delineate and describe climate discourses at different scales and across different registers, ranging from the local to the international, as assemblages of historically resonant and culturally salient themes, motifs, and symbols that are galvanized for specific political ends. These discourses do not always align, resort to differing rhetorical techniques, and overlap in their focus and goals. More often they appear to contradict each other, offering partial perspectives, and are characterized by omissions and selective emphases. In this chapter, I aim to examine these contradictions and silences that are partly rooted in the specific trajectory of the climate and, more broadly, in the environmental policies of the Indian government but that are also a reflection of the limitations and peculiarities of the currently ascendant political ideology. Understanding of climate change is shaped by cultural, religious, and historical lenses that are utilized to make legible a complex present in which economic, geopolitical, and political factors are deeply intertwined. India's climate policy, for instance, has remained consistent in several ways despite having shifted over time. It retains a significant element of flexibility that allows maneuvering room to take advantage of a changing international political-economic context. While the policy makers continue to position India in a way that is favorable to achieving its economic and strategic objectives, that positioning in itself has undergone changes that are consonant with the shifting international environment and the domestic political exigencies.

---

1. The Hindu notion of Kaliyuga has been compared to the scientific concept of the Anthropocene. Kaliyuga is described in the Hindu scriptures as a calamitous period marked by dissolute and weak humanity, similar to the Anthropocene, which is distinguished by environmental and economic profligacy. Amy Allocco (2014), drawing on ancient Hindu scriptures called *puranas*, describes Kaliyuga as a cosmological period of social-ecological crisis triggered by moral corruption: "These narrations foretell a housing shortage, a shrinking number of trees, and a diminished food supply and conjure an earth that will be whipped by harsh winds, scorched by intense heat, frozen by extreme cold, lashed by torrential rain, and parched from lack of water" (190). Nugteren (2005) ascribes to kaliyuga a certain defeatist attitude, a fatalism, especially in relation to ecological issues.

Amid dissonance and the seeming stochasticity of climate discourses proliferating at different scales—ranging from the neighborhood to the international scale—it is possible to discern certain salient tendencies and affinities. A trope commonly deployed by the middle-class householder and by Prime Minister Narendra Modi alike is that climate change is a manifestation of a world that has become skewed, not simply in the physical sense (e.g., unprecedented levels of greenhouse gas concentrations) but also in terms of its moral equilibrium. The only way to restore the moral order, according to both, is by recovering the lost body of knowledge and wisdom—deriving from *dharmic* traditions—and restoring it to its rightful place as a singular source of enlightenment. Another common strand that runs through otherwise disparate discourses is the pervasive substitution of rhetoric and symbols for substance and action. Yet another motif common to personal and official discourses, especially since 2014, is the almost obsessive quest for status and recognition by others ("image"), mainly internationally. Environmental issues and climate change are apprehended through mythic referents and cultural lenses forged in a world saturated with postcolonial grievances and globalized aspirations.

At the community level, the discourse of climate change is refracted through a multitude of quotidian concerns and questions. This study uses data about environmental values, perceptions, and practices of urban residents of a middle-class neighborhood in Delhi, India, collected over the past ten years. Environmental issues like air and water pollution feature prominently in discussions of quality of life. Increasingly, climate change or, more broadly, environmental change (*paryavaran mein badlav*) has emerged as a recurring theme as residents attempt to come to terms with the increasing frequency and intensity of extreme weather events. How do the residents make sense of these environmental changes that they perceive as taking place? How do they explain them? In other words, what kinds of explanatory frameworks are deployed to understand and perhaps assuage the fears and anxieties resulting from the perceptions of environmental deterioration? Except for perhaps groups of scientists and other experts working in research or policy, the environment is not an autonomous domain but rather a derivative one, refracted through myriad quotidian concerns.

Environmental discourses are constitutive of political and cultural anxieties, apprehensions, and aspirations and reflect worldviews that situate the subject in relation to its milieu. A dominant strand of Indian environmentalism, for example, has been described as antimodernist in its orientation: "Collectively, this environmentalism saw environmental degradation essentially as a result of the imposition of a Western, colonial civilization over a rooted, indigenous, Indian culture. In this form of environmental critique the conservation of nature was seen as dependent on the conservation of a past, pure, ideal nation, and its ancient culture and tradition—which, it was argued, was always in accord with the laws of nature. Ideas of nation and nature intermesh so completely in this view that defending the environment from pollution is virtually the same as defending the nation and vice versa" (Sharma 2012, 231).

Are these traditional/Western binaries invoked by the middle-class residents of Delhi? I argue that multiple traditions are being reimagined and redeployed for different purposes—moral and political—at different scales. The categories of native/foreign or past/present are, however, far from static; for instance, tradition is constantly being reinvented for political purposes and deployed for specific social ends. In what specific ways do the readings of tradition combine with the consumerist values of a neoliberal society to foreclose—or, alternatively, to open up—political possibilities? The state in India has had its own tradition of viewing climate change in the context of the country's place in the international political economy, which has undergone substantive upheaval since the 1990s. Popular attitudes and perspectives, however, are assemblages of discursive elements drawn from ideologies ranging from neoliberalism to romanticism pertaining to India's mythic past.

Adger (2001) describes the process by which actors utilize specific discourses to construct their identities. As opposed to being prefigured and given, identities are strategic and constitutive of individual and collective interests configured at the intersection of various emergent binaries, such as those encapsulated by state-society and individual-family relations. The intersection of environmental conservation and nationalism, with the category of nation being predicated on an essentialized relationship between people and places, religion and landscape, and society and morality, has fueled the Indian imagination since colonial times. In juxtaposing nature and society, a certain essential similarity, or homology, is assumed between the social order, characteristic of ancient India, and nature (*prakriti*), with social relationships and hierarchy being naturalized (Chapple and Tucker 2000). It is this assumed organic unity between society and nature and its putative rupture—the healing of which has been cloaked in terms of a call for revivalism—that has led to a deepening alliance between the movement for environmental conservation and right-wing political forces in the country. In Sharma's (2012) words, "Authoritarian moralism is a clear area of convergence between Green and Saffron in India, where environmentalism allies with feudal attitudes to instill moral codes, rules, and a discipline, which are hegemonically justified as vital to the life of the nation. This preference for a community of moral purpose represents an imposition of a moral orthodoxy" (261).

When the environment is viewed as a repository of not merely resources but also of culture and identity, environmental degradation becomes a moral-religious and military-strategic problem. The positing of the nation as a mother goddess (*Bharat Mata*) and the mapping of its terrain as a set of interlinked sacred spaces means that feelings of nationalistic pride and honor can be activated by conjuring external or internal threats. The alarmist element in the climate change discourse in India, and even India's most recent posture on climate change (both described in the next section on the international stage are aimed at generating a certain affect: fear and dread in the case of the former, and pride and self-assertion in the case of the latter. Even more importantly perhaps, the affect (or emotional gratification) is the goal, not the means to an end, as reflected in the lack of concomitant policy formulation and implementation. According to Sharma (2012), "In this form of environmental critique the conservation of nature was seen as dependent on the conservation of a past, pure, ideal nation, and its ancient culture and tradition—which, it was argued, was always in accord with the laws of nature. Ideas of nation and nature intermesh so completely in this view that defending the environment from pollution is virtually the same as defending the nation and vice versa" (231).

Climate change is thus read as another sign of a rupture of the essential unity of man and nature, with religiously informed tradition fostering harmony. As nature comes to be viewed through a web of immutable religious meanings, any change appears to be deviation or corruption, which can be rectified by restoring the status quo ante. In these security-centered nationalistic narratives, unity is emphasized, and any dissent is likely to be viewed as an attack on the sanctity of the mother-nation and, therefore, considered sedition. It is not difficult to see how framing of the environment, and specifically climate, in geopolitical terms can pave the way for top-down, undemocratic, and authoritarian approaches. In this worldview, the internal fissures and inequities in society and calls to address them are not only irrelevant to achieving environmental improvements but also antinational. According to Sharma (2012), "The past, it is argued, provides a record neither of harmony and utopia nor of the despoliation of nature by people. A valorization of tradition sits uneasily with the quest for justice" (254).

Another kind of "national" tradition—of India as an ex-leader and legatee of the Non-Aligned Movement (NAM)—is reflected in India's position in climate change negotiations. Indian leaders, beginning with Jawaharlal Nehru, independent India's first prime minister, have tried to take a moral stance on the global stage. This stance was rooted in the conviction that India's civilizational heritage affords it a unique place in the community of nations. This moral positioning was also grounded in the postcolonial politics of statism, rejection of capitalism, and Third World unity. The underlying philosophy of India's position was to make morality and ethics the basis of international relations. Often dismissed as moral hectoring or pointless petulance by some in the West, India's official pronouncements took the high road of universalist philosophy. In the post-1991 world, India's foreign policy began to shift in tone and tenor, with a marked embrace of the "pragmatic" or realpolitik in foreign affairs and a distancing from what was coming to be viewed as an impractical approach or even empty posturing. In the present political dispensation headed by Prime Minister Modi, the national foreign policy has unambiguously embraced international power politics. This shift has meant consideration of climate change not as an artifact of global political-economic inequities but rather as a negotiable item in the suite of issues confronting the country. The latter reconceptualization moves climate change from the moral field to the realm of diplomatic maneuvering and give and take. However, it does not necessarily imply that climate issues, especially greenhouse gas reduction targets, move from the symbolic domain to an arena governed by rationality. Far from it, India has continued to use moral-symbolic logic to bolster its case for recognition of its unique historical, social, and economic challenges in determining the country's obligations.

Climate Change and Culture: Degradation and the Need for Moral Therapy

I have carried out research related to environmental perceptions in a middle-class neighborhood of Delhi for more than a decade. In this section, I describe the ways in which climate change is understood and envisioned through the lens of social and cultural tropes that reflect the worldviews of the residents. When asked about climate change, some residents of the neighborhood react with a mixture of fatalism and exasperation—fatalism at the seeming downward spiral in which society appears to be caught and exasperation at the apparent inevitability of it. Nevertheless, despite consensus on the general decline in the environment, especially air and water pollution, most of the residents I interviewed demonstrated ambivalence and refrain from taking a categorical stance on the deterioration of the environment. Climate change is embedded in the larger discourse of decline and revival, fatalism and hope, and practice and policy. When asked what he thought about climate change and its impacts, the president of the neighborhood resident welfare association said:

> We keep hearing about climate change . . . in newspapers and on television. We are certainly affected by it. In the summer, we had so many dust storms that my mango crop in the village was damaged. Here, we have a shortage of water, although in this area, we are fortunate in having good water supply, but people in many other areas of Delhi are suffering. There is no doubt in my mind that the environment is worsening, but what can we do? People do not want to do their part—there is so much wastage of water and littering is a big problem. A lot of problems could be solved by population control, but look at how Delhi has grown. We don't even know anymore where it starts and where it ends. This government has done many good things, but it cannot do everything.

When prompted about the "good things" that the government has done, the interviewee listed fighting corruption and doing what is "good for the country" as the two accomplishments of the Modi government. Although climate change is clearly perceived and its effects felt, the issue is viewed with a degree of fatalism. None of the interviewees, for instance, mentioned India's position on climate change or the ongoing international negotiations to counter it, in which India has taken a leading role. For most people, climate change is refracted through a variety of political and social concerns, such as poor governance, corruption, and supposed abdication of individual responsibility. Another interviewee, a middle-aged officer in a public-sector bank, described the state of the environment like this:

> In the past few years, pollution has increased greatly. During the winter, there is so much fog that hardly anything is visible. The pollution starts with Diwali, when people burst firecrackers, with the result that for two days after [in] Delhi, the air is almost unbreathable. Then there are farmers who burn paddy stubble. The government banned firecrackers last year, and the pollution was not as bad as before. But the environment is changing, climate is changing—

there is no doubt about it. It is as if humans are getting punished for their *paap* [sins]. But being humans, our perception is limited, and we do not understand these things until it is too late.

This excerpt makes it clear that there is acute awareness of the pollution problems and environmental and climate change more broadly and their linkages with human actions. There is pervasive acknowledgment of the environmental problems and a desire for clean air and water and "green" neighborhoods. In terms of action or policies aimed at mitigation, however, there was near consensus that environmental and climate change was part of long-term decline. There was a strong tendency to view environmental degradation as part of a moral crisis emanating from certain historical events and forces.

Another interesting dimension of the interviewees' environmental perceptions is that they appeared to view a good environment (e.g., clean air and water, absence of pollution) not as a public good or entitlement to which everyone should have a right but rather as a privilege to which access is, unproblematically, a function of class and social status. Perhaps as a result of the percolation of the neoliberal ethos of individualism or because of entrenched hierarchical attitudes in Indian society, *environment* is reduced to amenities with skewed availability that is merely shrugged off as a social fact. When asked why climate change is not a priority for the government, another interviewee, a forty-five-year-old bank employee, said, "What can we expect from our governments when the politicians that are elected are mainly concerned with lining their pockets. They get elected through a politics of vote-banks and then spend their time in office making sure they get reelected. They are not interested in problems like climate change that require foresight [*doordarshita*]. This new government [the Bharatiya Janata Party (BJP) government that came into power in 2014] is sincere, but they have to clean up the mess of the past 70 years. It will not be easy."

There is also a certain degree of fatalism about the inevitability of the climate crisis that is rooted in a conception of human nature as flawed. A moral universe is constructed in which a paternalistic government tries to discipline a truant child, with, at best, mixed results. The apprehension of climate and environmental issues through social and religious-moral imagery and metaphors makes them at once legible while obscuring some of their important dimensions. The localization of the climate discourse and its disembedding from policy, ecological, and international contexts has many deleterious effects. The framing of climate and environmental problems in moral terms that require the imposition of discipline—nationalistic and individual—reduces the likelihood of a popular discursive breakthrough that would accord climate change the importance and urgency it deserves. In the next section, I will outline how this particular framing of climate change has been facilitated by a resurgence of cultural revivalism that seeks to subordinate science and technology to a constructed and mythic ancient past. Among other pernicious effects, one specifically stands out—the divergence and discordance between India's international position on climate change and its domestic contours.

Idealism vs. Muscular Pragmatism: Policy Reorientation as Cultural Assertion

This section situates Modi's emerging climate diplomacy in two contexts— domestic ideological reordering and resulting marginalization of the secular-scientific tradition—in addition to the reworking of India's foreign policy orientation. The imperatives of these two contexts often complement each other but at other times diverge and even collide, producing policy inaction and policy paralysis. The ascendance of the BJP, the main center-right political party in India, to power has reignited and energized debates that have long pedigrees in India. In the period leading up to the independence of India from British rule, the freedom struggle was spearheaded by the Congress party, which embraced a modernist vision for the country. In this view, the path ahead for India lay in vigorously adopting a "scientific temper," not just in areas of technological and economic development but also in the social and political spheres. Jawaharlal Nehru, India's first prime minister and a stalwart of the country's freedom struggle, placed particular emphasis on social reform and on purging society of regressive and superstitious practices. In this dominant perspective, tradition needed to be subjected to critical scrutiny and recast in light of the modern-day imperatives of growth, equity, and justice. Internationally, Nehruvian policies rejected power politics and realpolitik and, instead, advocated a policy of reconciliation and cooperation based on the idea that relations between nations need not be a zero-sum game. Nehru was a founding member of the NAM, which sought to bring together the erstwhile colonized nations on the same platform of equidistance from both power groupings of the time: the Western nations and the Soviet bloc.

With the collapse of the Soviet Union in 1991, India suddenly found itself in a unipolar world for which it was unprepared. Coupled with a sense of political isolation, the economic crisis that occurred forced India to liberalize its economy, leading to privatization, partial convertibility of the rupee, and lowering of import tariffs. With the unleashing of the twin forces of liberalization and globalization, the rise of the middle classes was catalyzed—a development feted by international economic institutions such as the International Monetary Fund and the World Bank. In the domestic realm, however, the global economic integration and concomitant ascendance of consumerism considerably shrunk the ideological space for progressive politics. The critique of Nehruvian economic and foreign policies as woolly-headed idealism, detached from Indian realities and tradition, began to gain traction with a middle class hungry for international status. Although the anti-Nehruvian camp in Indian politics had existed since before independence, the rejection of the progressive ideals it represented has gone much farther in the past two decades. On the foreign policy front, the liberal internationalism that gave birth to the NAM has been substituted by the supposedly cold calculus of national interest, grounded in clear-eyed assessment of global power relations and India's strengths and weaknesses.

The changes in India's foreign policy that began in 1991 and picked up pace in the past decade include abandonment of the goal of a multipolar world and an ever more explicit tilt toward the United States. It is interesting that this change in India's foreign policy orientation, which is presented as an embrace of "realism" and "empiricism," is often justified by rhetorical appeals to the genius of Chanakya, a fourth-century BC political philosopher (Sullivan 2014). Chanakya, sometimes described as the Indian Machiavelli, is invoked as a hard-headed, pragmatic thinker whose ideas stand in contrast with the putative romanticism of India's foreign policy in the half century following its independence. What is notable about the remarkable resurgence of diplomatic, political, and popular interest in Chanakya is the nearly complete absence of discussions about his peculiarly amoral vision of society and an uncritical celebration of tactics and strategies, alluding to the ascendance of a mechanical view of the polity and of human nature (Liebig 2013).

In the domestic realm, the repudiation of the Nehruvian social vision has resulted in an aggressive celebration of tradition as a panacea to contemporary societal problems. This "return of tradition" encompasses a revivalism in phenomena as disparate as Ayurveda and conversion of Muslims to Hinduism (in a ritual described as *ghar vapasi*, or homecoming). A particular strand of this antimodern mindset comprises glorification and invocation of the greatness of ancient India and its supposed achievements in science, technology, and medicine. Prime Minister Modi, for instance, while addressing a public gathering, reminded the audience that plastic surgery and genetic medicine had existed in ancient India and that the task before the nation was to recover its lost greatness (Maseeh Rahman 2014). The narrative of civilizational decline is that India was once a fountainhead of scientific and mathematical knowledge and wisdom until the decline began as a result, at least in part, of foreign (read "Islamic") invasions. This mindset of uncritical glorification of a mythic past has been described by Nanda (2015): "Indocentrism . . . has been the hallmark of Indian historiography of science. Like the mirror image of Eurocentrism, which holds the 'Greek Miracle' to be the original source of all sciences, Indocentrism holds ancient and classical (that is, pre-Islamic) India to be the Givers and all other civilisations to be eager and grateful Receivers. If an idea can be found in India and in some other place in a comparable time frame, our Indocentric historians simply assume that it must have travelled there from India, but never to India."

Since 2014, the national politics has been dominated by the right-wing Hindutva ideology, the socioeconomic and cultural effects of which have been sweeping. This shift has led to significant change in India's policies in international relations and its stance on the global stage. Historically, India's position on climate change was marked by a certain defensiveness aimed largely at protecting its autonomy in the political-economic realm.[2] This position was replaced by an extroverted thrust fueled by a desire for India to assume its rightful place as a *vishwaguru*[3] in the comity of nations (Hall 2017). It has been noted that despite an ambitious-sounding agenda of the current right-wing political dispensation, aimed at transforming Indian foreign policy, it is in "only two areas [that] there is a more obvious connection to Hindu nationalist or Vivekanandan ideas: in Modi's treatment of climate change and the global environment, and in the role that religion and spirituality should play in mitigating international and civil conflict, including terrorism" (Hall 2017).

In the case of climate change, arising from India's position as the second-largest global emitter of greenhouse gases and its great vulnerability

---

2. Following its independence from Britain in 1947, and in continuation of its anticolonial stance, India pursued policies of economic autarky aimed at creating self-sufficiency and reducing dependence on foreign powers.

3. The word *vishwaguru* can be roughly translated as "world leader" in a political, economic, and, most importantly, spiritual sense.

to climate change, Modi senses an opportunity to catapult India into the ranks of the foremost nations on the world stage. It is interesting that Modi has hyphenated climate change with terrorism as two of the most pressing issues confronting the international community today ("At Davos, PM Modi Hits Out at Protectionism" 2018). The common thread between these two problems in Modi's narrative is that India is both the victim of and a potential solution to these thorny problems. In laying out the connection between climate change and terrorism, Modi invoked the concept of *vasudhev kutumbakam* (all the world is one family) and points to these issues as preventing the emergence of a truly global consciousness. Typically, the resort to ancient Indian philosophical and mythological traditions and emphasis on multiculturalism and inclusive growth are recurring themes in Modi's speeches abroad.

The resort to symbolism and appeals to nationalism and cultural pride when the issue of hard choices to tackle climate change comes up has become utterly routinized in Prime Minister Modi's political theater. The use of symbolism, as in the invocation of the ancient Hindu texts, *Vedas*, is designed to exert rhetorical leverage both internationally and domestically. It is interesting that the use of hallowed cultural referents as a substitute for policy capacity and effectiveness is not just limited to the climate arena. Most recently, in response to the COVID-19 pandemic, Modi announced a nationwide lockdown. Additionally, however, he exhorted the citizens to light *diyas* (earthen oil lamps) on the evening of April 12, 2020. Invoking the symbolism of Hindu festival of Diwali, when *diyas* are lit to mark the victory of good over evil, Modi tweeted that "salute to the light of the lamp which brings auspiciousness, health and prosperity, which destroys negative feelings." This was Modi's second symbolic onslaught on COVID; earlier he had called for clapping, beating of plates and ringing of bells on March 22, 2020, at 5 p.m., to show appreciation for doctors and other health professionals battling COVID. This had elicited a caustic response from many, including Shashi Tharoor, the member of parliament from the opposition Congress Party, who proclaimed that "clapping doesn't kill the virus" ("Clapping Doesn't Kill Virus" 2020).

Another example of Modi's attempt to link climate change to Indian culture and history on the world stage was his espousal of the merits of yoga in finding solutions to climate change (Adler 2014). Observers were left scratching their heads at the reference to yoga; however, Modi had in mind this:

We can achieve the same level of development, prosperity and well-being without necessarily going down the path of reckless consumption.... We treat nature's bounties as sacred. Yoga is an invaluable gift of our ancient tradition. Yoga embodies unity of mind and body; thought and action; restraint and fulfillment; harmony between man and nature; a holistic approach to health and well-being. It is not about exercise but to discover the sense of oneness with yourself, the world and the nature. By changing our lifestyle and creating consciousness, it can help us deal with climate change. Let us work towards adopting an International Yoga Day.

The excerpt above is from the speech that Modi delivered to the annual session of United Nations General Assembly (soon after being elected prime minister for the first time). He skipped the World Climate Summit a few weeks earlier, causing much consternation among the international climate community. In addition, the rest of his speech reiterated the main points of India's long-held position on climate change—namely, that the principal responsibility for mitigating climate change lay with developed nations. It is easy to see why Modi skipped the climate meeting where the discussion would have been technical, involving emissions reduction targets and obligations, whereas the General Assembly provided the right forum for a freewheeling address complete with rhetorical flourishes.

Modi's formulaic shifting of the discursive parameters to a plane defined in terms of abstract and affective albeit culturally resonant imagery did not escape critical attention: "Frugality, resourcefulness, worshipping trees et al are some of those ideas he tries to reframe and re-apply to the modern world with very little practical sense. The Hindu mores and traditions offer values and principles to base our lives on but don't drive policies for the 21st century challenges that have emerged from a very different reality" (Kumar 2015). More charitable commentators have termed this discursive sleight of hand as part of a quest in which a "rising India needs a new normative agenda" as part of an enduring "posture built into Indian foreign policy [reflecting] an early aspiration of exceptionalism through moral leadership" (Sengupta 2019). Dismayingly, although perhaps expectedly, the Modi government's domestic policies—whether environmental or social—have not conformed with the rhetoric deployed internationally. Since 2014, a divisive social agenda that amplifies religious differences between Hindus and others has been put into practice. Contrary to pious proclamations of the world being a family, intolerance—especially the religious kind—has become a prominent feature in the public spaces. Incidents of mob violence, particularly led by *gau-rakshaks* (cow protectors) targeting people accused of cow slaughter (mostly Muslims) have increased manifold. Similarly, the past four years have seen increasing intolerance of dissent aimed at government and corporate-led development projects ("From Democracy to Mobocracy" 2018). The Modi government, whose proximity to corporations is no secret, has directly and blatantly pursued economic policies that have benefited big business, contradicting its slogan of *sabka saath, sabka vikas* (with all, development for all). Some commentators see disturbing parallels in the rise of the Modi regime with other politically right-wing movements in Europe and America and trace these developments to increasing inequality, the decline of the organized left, and a resort to culturally regressive themes and symbols as conduits for popular resistance ("Terror, by Another Name" 2018; Mukherji 2018).

The Modi government's environmental policies at home have been notably lax in enforcing the laws and have consistently placed economic interests—or, more accurately, narrow corporate interests—ahead of public health and environmental concerns. In the past several years, deterioration in urban air quality in large metropolitan areas, especially in the capital city of Delhi, has attracted considerable scrutiny and triggered anxious media coverage. Amid rising public concern about the adverse health impacts of poor air, it has not gone unnoticed that these changes are unfolding against the backdrop of a government that clearly favors development over the environment (Gettleman, Schultz, and Kumar 2017). Notwithstanding the rhetoric about protecting the environment, the BJP government's environmental policies have worked at cross-purposes to its stated lofty goals: "The conference, where Modi wielded his environmental rhetoric, is expected to result in the dilution of major environmental policies to give a free run for industry. Doing away with environmental clearances, revoking the rights of tribals to forest land, approving coal mining in dense forests; the list of policy changes the Government introduced has capital, not the environment or our traditional values, at the heart of it" (Kumar 2015).

The contradictions between the supposed ancient Indian tradition of respect for the environment and the antienvironmental policies have manifested themselves in myriad ways. In October 2018, Prime Minister Modi's signature environmental campaign for Ganga cleanup came under intense scrutiny and broad criticism. A noted scientist-environmentalist G.D. Agrawal died while undertaking an indefinite fast in support of his demands aimed at stopping construction of hydropower projects on the Ganga River. Agarwal was a well-known and publicly admired figure who represented a rare combination of scientific expertise and saintly virtue. Exemplifying the Gandhian social idiom of self-sacrifice, he led an austere life devoted to a single-minded pursuit of "saving" the Ganga River from power projects that would have interrupted the river's free flow. He was described like this in an obituary: "G.D. epitomized simple living and high thinking. A greater part of his last 25 years was spent in Chitrakoot, where he became an honorary professor at the Mahatma Gandhi Grameen Vishwavidyalaya. This eminent scientist swept his own floors, washed his clothes and cooked his meals in a spartan 200-sq ft cottage. A bicycle, an ordinary state transport bus or second-class train compartment were his preferred modes of transportation" (Chopra 2018).

The much-ballyhooed combination of qualities—scientific and professional rigor couched in a saintly aura—would have been expected to endear G.D. Agrawal to the ruling party. After all, who could be a better icon for the synthesis of Indian cultural ideals and scientific expertise? However, despite writing a letter to Prime Minister Modi with four demands aimed at preventing hydropower development on Ganga, he received no response. On October 11, 2018, he died on the 112th day of his fast-unto-death "seeking effective action from the Government of India for the conservation and protection of the Ganga" (Chopra 2018). The Indian government's flagship plan for Ganga cleanup, called the "Ganga Action Plan," has been plagued by inadequate resources, poor institutional functioning, and overall shoddy implementation, providing another example of the yawning gap between tall promises and poor results (Alley 2016).

The trend of Modi invoking lofty notions of universal brotherhood and the shared global imperative for environmental stewardship while pursuing almost diametrically opposed policies continues unabated. If anything, the pandemic-induced economic crisis in India has led the BJP government to drop nearly all pretense of environmental safeguards and oversight. In May 2020, the environment ministry cleared dozens of proposals for mining and other extractive activities in sensitive ecological areas after a cursory review in which its own guidelines for systematic expert review were openly flouted (Gokhale 2020).

India and Climate Change: Policy and Performance on the International Stage

Over the past decade, climate change has taken center stage in the arena of international relations. Because it is never merely a domestic issue, climate change features critically in India's international image. The consequences of climate change have been predicted to be particularly dire for developing countries, as most faced with resource scarcity, governance problems, and adverse environmental conditions. In India, scientists and others have warned about the deleterious socioeconomic effects of climate change that are already beginning to appear. Climate change presents an objective and imminent threat to India's economy and stability; however, it also provides a platform to political and policy entrepreneurs projecting a nationalist agenda. In the next sections, an overview of India's complex engagement with international climate change negotiations and policies is presented. India's climate positioning, while often overtly political and evolving, is targeted to appeal to both international and domestic audiences.

Global negotiations about climate change, primarily between developing and developed countries, have long been caught up in divisive and acrimonious debates about history, equity, and culpability. These negotiations have come to hinge on the role of finance and technology, with equity concerns taking a backseat. Although the developing countries have often demanded direct fund and technology transfers, the developed countries (especially the European Union) have emphasized the role of markets in adapting to climate change. Nevertheless, the efficacy of markets and the capacity of the developing countries to efficiently use funds, given inadequate governance structures, remain dubious (Saran 2010).

In tandem with global geopolitical and ideological shifts, the Indian state's response to climate change has evolved considerably over time, although it has been suggested that the change falls considerably short of transformative (Miller and de Estrada 2018). In the 1990s, India adopted what is described as the "climate justice" position. This stance entailed a critique of the notion of national emissions reduction quotas. Most developing countries, especially India, took the position that national targets should take into account the practices—both historical and contemporary—of the Western countries that contributed massively and disproportionately to greenhouse gas emissions. In other words, a climate policy that adopts an ahistorical stance of uniform national emission reduction targets, while neglecting the historical dimension of the problem, would exacerbate global economic inequality by severely and unfairly penalizing developing countries. A possible resolution to the problem of determining emission reduction targets lay in the approach outlined by Agarwal and Narain (1991), who envisioned a per-capita emission allowance. This approach would support developing countries because of their large populations.

In the 2000s, India's official position on emissions reduction began to shift. A key figure who articulated the turning point was the country's environment minister Jairam Ramesh. In the run up to the climate negotiations in Copenhagen in 2009, Ramesh announced an important change in India's stance. He argued for binding international agreements on emissions reduction, which marked a first in terms of India's willingness to agree to specific emissions reduction targets. Manjari Chatterjee Miller and Kate Sullivan de Estrada (2018) argue that the goals of India's policy makers have been remarkably consistent—adherence to an approach that envisions "common but differentiated responsibilities and respective capabilities" (45)—despite seeming variations in the official climate discourse. According to Miller and de Estrada, the primary objective has been to ensure that binding emissions reductions do not come at the expense of economic development in India. In other words, given India's great need for cheap energy options (e.g., coal), any agreement that makes this goal difficult to achieve is viewed as constraining the country's "development space" and, as a result, is considered a nonstarter. However, as nonrenewable energy technologies have developed, India has made its support for emissions reduction conditional on the availability of financial support and technology transfer.

India ratified the Paris Agreement on October 2, 2016. The symbolism of this date, Mahatma Gandhi's birthday, is unmistakable for this important milestone in India's climate policy—namely, that India remains committed to Gandhi's global vision of humanity embracing a simple and frugal lifestyle and respecting ecological limits. Under the terms reached, India agreed to meet at least 40 percent of its energy needs from renewable resources by 2030. However, India's proposals to bring about the agreed-upon reduction in the carbon intensity were noted as being heavy on symbolism ("India's ratification talks extensively of Gandhi and Yoga") and lacking in specifics (Vyas 2016). India's ratification of the Paris Agreement came with strings attached, which underscored continuity with its long-standing position about striking a balance between climate and development. In particular, India pushed for an annual $100 billion fund transfer from developed to developing countries to facilitate transition to green technologies and vigorously opposed inclusion of social and environmental safeguards in the World Bank–funded projects (Mehra 2015). Considering the differences in

Modi's approach compared with his predecessors, a commentator on international affairs noted approvingly, "After decades of Indian defensiveness at international forums, Modi is signaling a new strategy that plays hardball on India's core national interests, demonstrates tactical flexibility, avoids ideological argumentation, builds new coalitions, and contributes to positive outcomes" (Mohan 2015).

In summary, India's forays into international climate politics can be located on an evolutionary arc shaped by global and domestic pressures. From viewing climate as a problem to be tackled by diplomacy in the 1990s to considering it as an obligation and even an opportunity a decade later involves a substantial journey. Navroz Dubash and colleagues (2018) describe "the slow dissolution of the wedge between climate as a diplomatic versus a developmental issue [as] best characteriz[ing] India's changing role in international climate discussions" (397). The domestic discourse that had largely been subservient to India's international position of making per-capita emissions the lynchpin of efforts to allocate emissions reduction targets also began to shift. India had painted a dichotomous picture between developed and developing countries based on the huge gap in their per-capita emissions and used this approach to pin the responsibility for reducing emissions on rich countries. Increasingly, however, intracountry emissions disparity began to undermine India's position—which had been bolstered using the notion of sovereignty. "The charge was that the high emissions of India's wealthy population were camouflaged by the very low emissions of India's poor, and that notions of equity should more properly include intracountry equity" (Dubash et al. 2018, 398). Over time, India's engagement with climate policy has deepened without doubt, but "given the overhang of immediate development challenges, climate change can only be salient to politics and governance if a robust analytical framework is developed to integrate climate considerations alongside and interwoven with pressing development challenges" (Dubash et al. 2018, 415).

Environmental discourses are not totalizing, that is, they do not foreclose multiple and even contradictory meanings, and actors actively reinterpret, modify, and deploy discourses and symbols in pursuit of specific objectives (Dryzek 2005). In India, as noted, there is a long history of climate change being viewed as a problem of international relations, with policy makers viewing their primary aim as avoiding the constraints on growth sought by Western powers. Domestic policies aimed at ameliorating climate change have lagged far behind the geopolitical discourse surrounding its international dimensions. It has been noted that in Indian policy circles, the pressure to respond to international agreements is felt more acutely than the urgency to take adaptive measures domestically (Dubash 2012). This observation implies the existence of a yawning gap between what is professed at the international forums and what is practiced domestically.

The contradiction between grandiose pronouncements and commitments made on the international stage and the relative lack of action on the ground has been noted (Sivaram 2017). The state of Gujarat, of which Modi was the chief minister for more than ten years, is emblematic of how climate change adaptation policies and practices are captive to hard political-economic exigencies. Under Prime Minister Modi, environmental discourse, including climate discourse, has been infused with religious and cultural symbols, almost as a substitute for concrete action and tough tradeoffs. Notwithstanding Modi's lofty assertions, the implausibility of a conservative prime minister taking a proactive position on climate abatement—even as a departure from India's traditional position on the issue—has not escaped attention (Chemnick 2016). Since his tenure as chief minister of the Indian state of Gujarat, Modi has demonstrated his ambition to make a mark on the international stage. His climate adaptation policies in the state of Gujarat, however, were premised on a narrowly technical and market-based ideology that elevates expert knowledge and economic rationality above meaningful community participation and traditional knowledge (Venkatasubramanian 2016). Two larger points about Gujarat's piecemeal attempts at climate change are widely applicable. First, climate change as a discourse is a loose agglomeration of heterogeneous economic, social, and political agenda items and can be mobilized by a variety of social actors for their own ends. Second, the lack of synergy among economic, social, and climate policies is a major constraint to climate adaptation (Venkatasubramanian 2016).

Indian Environmentalism and Hindutva: Putting the Saffron in Green

Environmentalism is generally conceived as a collective movement that aims to stimulate, through direct and indirect action, policies and practices favorable to the environment. In India, environmentalism does not have a monolithic structure: it is practiced and reflected through the actions, policies, and ideas of a variety of actors and institutions such as the state; nongovernmental organizations, including overtly environmental organizations; religious and social bodies; industry; and local communities. Climate change—perhaps the most pressing environmental issue of all time—is perceived and responded to in diverse and divergent ways. As the preceding sections have shown, scale plays a critical role in framing and reframing the issue, with the community-level and international discourses of climate change constructed in very different ways. In this section, I turn to the role of ideology and, especially, recent ideological shifts in India's body politic that have come to permeate all levels of climate discourse, rhetoric and, to a lesser extent, policy. The movement of Hindutva from the fringes of Indian political life to its center has manifold ramifications for the social contract that binds the individual and the state in a web of reciprocal relations. The next section is devoted to the explication of the ideology of Hindutva and the reconfiguration of social, economic, and political lives of the citizenry it envisions and its implications for the human–environment relationship.

As an ideology, Hindutva comprises "a highly structured belief system involving the interpretation of the past, an analysis of the present, and a set of precepts and imperatives for the future conduct" (Banerjee 1991). As an all-encompassing and ambitious political project, it seeks a moral reordering and reimagining of the social and political universe. In 2014, India's national elections led to the victory of the BJP and the elevation of Narendra Modi as India's prime minister. With an unabashed Hindutva-based social and political agenda, it was the first time that a cultural-nationalist discourse occupied a central place in India's political life. Gone was the relatively discreet and sporadically apologetic ideological approach of the previous BJP government that had governed the country in mid-1990s. A cultural-nationalist populism has been articulated and gained considerable traction in making a political subject of "the people," defined as excluding minorities, especially Muslims, and left-liberal "elites." Chakravarty and Sharma (2020) describe the indispensability of a charismatic and authoritarian figure for this attempt to fashion Hindus into an ethnonational community: "In addition to delimiting the authentic 'people,' this form of populism typically relies on a leader who claims to be the sole representative of the people and the embodiment and authority of the popular will. Modi is a paradigmatic example of such a leader."

The Modi regime has openly professed its majoritarian agenda with a scope that ranges from revising the school curriculum to instilling the correct patriotic values in children and shaping a muscular projection of India in its region and, most important, on the global stage.

The majoritarian thrust of the government has made some commentators pessimistic about the future prospects of India's civilizational ethos—threatened as they are by the unleashing of homogenizing forces (Jaffrelot 2018). One of the common threads between the domestic and international dimensions of Hindutva ideology is the belief that India (or, more accurately, the Indic civilization) was one of the greatest civilizations in the world and was systematically deprived of the recognition that it richly deserves. From this perspective, the achievements of the Indic civilization, especially ancient India, were eclipsed, if not entirely effaced, by successive "foreign" rulers. The period of foreign rule encompasses both the Islamic (1200–1857) and British (1857–1947) phases of India's history. The Hindutva project is strongly fueled by a sense of profound grievance and historical injury originating in the belief that Indian society and culture were actively suppressed by the "colonialism" imposed over a period of nearly one thousand years. With Modi coming to power, many feel that India has the first authentically Hindu leader, who will help India reclaim her lost glory and occupy her rightful place in the comity of nations.

One of the primary tasks that Hindutva ideologues have set for themselves includes furnishing an explanation for what they perceive as India's material and technological backwardness. How can India claim the mantle of an emerging global power in the face of its undeniable absence from the ranks of the world's advanced economies? Technology and science emerge as key battlegrounds on which these ideologues seek to assert a cultural and religious revivalist agenda. The tendency to draw on the supposed wisdom and knowledge of the ancient Hindu civilization has been critiqued for its tendency to romanticize and present a lopsided view of history (Nanda 2005). In addition to the factual basis of the claims being made, their relevance to contemporary social and environmental problems has also come into question. It can be argued that these appeals to the glories of a putative golden age are premised on "a sense of cultural crisis [that] connects the power of religion to the power of ecology" (Jenkins and Chapple 2011).

Revivalism does not mean that all tradition is treated in the same way and that all of the past is accorded equal recognition. The Hindutva movement has embarked on a quest to recover and resurrect "grand" tradition that can be used to project India's glory, power, and status. Traditions that do not fit in this category—namely, folk traditions based in environmental knowledge and practice or tribal traditions rooted in subsistence and autonomy—scarcely feature in the revivalist agenda. Another notable omission is the Gandhian tradition of rural autonomy based on ideas of simplicity, self-sufficiency, and communal harmony. Consequently, the role of radical environmental discourse in climate change debates in India has been marginalized. The prescription of Gandhian economics and the focus on the rural–urban and rich–poor disparities within India, which form the crux of the radical green discourse, have not received much consideration in policy circles.

The articulation and circulation of climate policy discourse in India is far from monolithic and variously draws on religious, moral, scientific, and economic vocabularies. These discourses connect and overlap in curious and contradictory ways, evoking the other socioecological moments and movements in Indian history in which similar contradictions were apparent. The role of religion in attenuating such contradictions is not new; the quest for locating scientific expertise in the realm of ancient myth and religion contravenes what Haluza-Delay, Veldman, and Szasz (2013) describe as the "paradigmatic barriers" that separate religion from science. Whereas scientific knowledge is premised on a separation of the material and nonmaterial realms, religious cosmologies weave these aspects together through rituals, doctrines, and everyday practice. In religions such as Hinduism and Buddhism, with no central authority and a lack of a unitary central structure, plurality prevails at various levels. Among Hindus, for instance, the attitudes toward pollution in the Ganga River are mediated by their beliefs about the nature of the object (i.e., the river). The transcendent qualities of Holy Ganga, as a deity, for many, render it impervious to physical pollution. As Alley (1994) notes, "Logically, Ganga and *gandagi* (pollution) are separate entities; they are not collapsed or dissolved, as scientific theory would have them, within a single measurable substance called water" (141). This statement reveals the cultural logic that puts the different dimensions of the same ostensible phenomenon, environmental pollution, on different planes by viewing them as fundamentally incommensurate.

The role of religion, especially Hinduism and Buddhism, in fostering environmental conservation has invited considerable scholarly scrutiny. The concept of *ahimsa* (nonviolence), which is important in Hindu and Buddhist religious traditions, has been credited with the environmental and prolife ethic practiced by many Indigenous communities throughout South Asia (Sivaramakrishnan 2015). In this context, the sacred groves that are found in many regions of South Asia have been extolled as a specific cultural practice that has contributed to the preservation of biodiversity (Nugteren 2005). More broadly, the relationship between culture and environment has been the subject of much conjecture and theorizing going back to ancient Indian scriptures. Drawing on Zimmermann's study (1987) of the Vedas, Barnes and Dove (2015) describe the association between different types of forests and civilizational attributes: "The ancient Vedic texts describe a cosmological divide between the semi-arid savanna (jangala) of western India and the perennially wet forests (anupa) of eastern India, which in turn is based on a fundamental underlying polarity between agni (fire) and soma (water). Like the climatic divides . . . this one has a normative dimension: one zone is the abode of the civilized Aryan, the other is that of the uncivilized barbarians" (Barnes and Dove 2015, 32). Further, Zimmerman (1987) writes, "In ancient India, all the values of civilization lay on the side of the jungle. The jangala incorporated land that was cultivated, healthy and open to Aryan colonization, while the barbarians were pushed back into the anupa, the insalubrious, impenetrable lands" (Barnes and Dove 2015, 18).

Indian environmental movements have differed in important ways from their Western counterparts. Often these struggles have been concerned with both conservation of the environment (e.g., forests) and the preservation of rural livelihoods. This was understandable because rural livelihoods in many areas depended on the integrity of the local environments. In a fragile ecosystem such as the Himalayas, the Chipko movement focused on ending commercial logging but also was an attempt to maintain rural residents' access to forest resources on which their livelihoods depended (Guha 2000). In mobilizing rural communities, movements like Chipko[4] have used religious metaphors and imagery extensively, mobilizing traditional social structure and building powerful political alliances.

When religious considerations are applied to a geographic and political entity such as the nation-state, the entity becomes a sacred space whose violators are not just aggressors but "the other" with genocidal intentions. The alarmist dimension of climate discourse in India taps into these anxieties.

---

4. Sacred groves in South Asia that harbor a significant diversity of flora and fauna and that often serve as biodiversity hotspots are another example of religiously infused conservation (Gupta, Kohli, and Ahluwalia 2015).

In some quarters, climate change has been framed in terms of its damaging implications for national security, perhaps to appeal to more conservative policy makers (Busby 2008). Climate change has been viewed as disturbing the status quo in South Asia, for instance, in leading to the displacement of tens of millions of people, especially in the low-lying nation of Bangladesh. India is presented as the likely victim of this displacement, and images of impending demographic invasion—and, along with it, sociopolitical unrest and economic disruption—are vividly painted, tapping into the latent and historically grounded fears of the "other" as Muslims (Narang 2017). For the Hindu nationalist government, the geopolitical framing of the climate change issue is particularly convenient because it harnesses the metaphor of the motherland under assault by unholy outsiders, which constitutes a pillar of the Hindutva movement.

Conclusion

Climate discourse in India is not monolithic but rather fragmented by discrete social, cultural, religious, and political elements that overlap in distinct configurations in different realms. Religious values and perspectives are insinuated into climate discourses in direct and indirect ways. Religious worldviews are often deployed in a straightforward manner in discussions, quotidian and otherwise, to affirm or challenge certain perspectives; however, religious, and specifically Hindu, symbols and motifs are strategically deployed to make or further certain political claims. The notion of *kaliyuga* in community discourses of climate change, for instance, is used to legitimize the status quo and a kind of fatalism in matters concerning the environment and climate.

On the international stage, India's religious and cultural traditions are invoked to buttress the country's claim to global leadership, or at least to a position of "first among equals" among the developing countries. When the prime minister proclaims "vasudhev kutumbakam," he asserts that global environmental stewardship is deeply rooted in his religion and culture and thereby "naturalizes" India's claims to a leading role in global environmental politics. This approach is partly designed to counter the accusation that India does not take its international obligations seriously and has been dragging its feet on binding greenhouse gas emissions reduction targets. Furthermore, it underscores the Indian state's attempts to project itself as a moral force in the global realm. Ultimately, it could be argued that these discursive techniques, in combination with the overt position of common but differentiated responsibilities, are aimed at carving out a policy space for India that would allow it to pursue economic and political policies that are congruent with its national interest.

The use of environmentalism discourses in appeals to recover a lost past have been documented in many different cases. Environmentalism—whether organized and overt or spontaneous and implicit—inevitably tends to be prescriptive, so the question of what environmental baseline should be restored emerges as a critical issue. Although all renderings of the past are selective and partial, some are more exclusionary, sectarian, and confrontational than others. The community framings of climate change juxtapose a dysfunctional present with an ideal past. We lament not just the loss of ecological integrity but also the erosion of a social and political order. The internal contradictions in the micropolitics of climate change, reflected in both community-level discourses and official pronouncements on the international stage, are fairly obvious. Both emphasize austerity, self-discipline, and a recovery of tradition—as in the use of kaliyuga as a diagnostic and yoga as a moral therapy—and yet openly embrace consumerism as a personal and policy goal. In fact, even yoga—the putative antidote to the rampant materialism of kaliyuga—is packaged and commodified, now by the state, to project India's soft power (Mazumdar 2018).

The injection of religious, specifically Hindu, symbols and themes in the environmental movement has the effect of mobilizing majoritarian constituencies with exclusionary social and political effects. This approach is not conducive to building a broad national consensus on climate policies, which risk becoming hostage to partisan politics. If effective climate policies are to be designed and constituted, the critical need is for a discourse that is scientifically robust, culturally salient, economically equitable, and politically inclusive—this cannot be overemphasized. In this endeavor, the potential of religion to positively shape and influence such discursive formation remains, for the present, deeply underutilized.

References

Adger, William. 2001. "Scales of Governance and Environmental Justice for adaptation and mitigation of climate change." *Journal of International Development* 13 (7): 921–31.

Adler, Ben. 2014. "Yoga Could Be an Answer to Climate Change, Says India's Prime Minister." Grist.org. Accessed October 2, 2014. https://grist.org/climate-energy/yoga -could-be-the-answer-to-climate-change-says-indias-prime-minister/.

Agarwal, Anil, and Sunita Narain. 1991. *Global Warming in an Unequal World: A Case of Environmental Colonialism*. New Delhi, India: Centre for Science and Environment.

Alley, Kelly. 1994. "Ganga and Gandagi: Interpretations of Pollution and Waste in Benares." *Ethnology* 33 (2): 127–45.

Alley, Kelly. 2016. "Rejuvenating Ganga: Challenges in Institutions, Technologies and Governance." *Tekton* 3 (1): 8–23.

Allocco, Amy. 2014. "Snakes in the Dark Age: Human Action, Karmic Retribution, and the Possibilities for Hindu Animal Ethics." In *Asian Perspectives on Animal Ethics: Rethinking the Non-human*, edited by Neil Dalal and Chloe Taylor, 179–201. Milton Park, UK: Taylor & Francis.

"At Davos, PM Modi Hits Out at Protectionism, Says Terrorism, Climate Change Grave Threat." 2018. *Indian Express*, January 23, 2018. Accessed July 27, 2021. https://indianexpress.com/article/india/narendra-modi-world-economic-forum-davos-climate-change-terrorism-5036197/.

Banerjee, Sumanta. 1991. "'Hindutva': Ideology and Social Psychology." *Economic and Political Weekly* 26 (3): 97–101.

Barnes, Jessica, and Michel Dove. 2015. *Climate Cultures: Anthropological Perspectives on Climate Change*. New Haven, CT: Yale University Press.

Busby, Joshua. 2008. "Who Cares about the Weather? Climate Change and US National Security." *Security Studies* 17 (3): 468–504.

Chakravarty, Udipta, and Rohit Sarma. 2020. "The Limits of Hindutva." *Jacobin*. Accessed May 12, 2020. https://www.jacobinmag.com/2020/01/hindutva-india-citizenship-amendment-bill-muslims-modi-bjp.

Chapple, Christopher, and Mary Tucker. 2000. "Hinduism and Ecology." Cambridge, MA: Harvard University Press.

Chemnick, Jean. 2016. "India's Conservative Prime Minister Proves Unlikely Climate Ambassador." *Scientific American*, June 10, 2016. Accessed July 4, 2021. https://www.scientificamerican.com/article/india-s-conservative-prime-minister-proves-unlikely-climate-ambassador/.

Chopra, Ravi. 2018. "Scientist and Tapasvi." *Indian Express*, October 16, 2018. Accessed July 27, 2021. https://indianexpress.com/article/opinion/columns/gd-agarwal-namami-gange-ganga-narendra-modi-5403520/.

"Clapping Doesn't Kill Virus: Shashi Tharoor on Modi's Janata Curfew." 2020. *Economic Times*, March 22, 2020. Accessed July 27, 2021. https://economictimes.indiatimes.com/news/politics-and-nation/clapping-doesnt-kill-virus-shashi-tharoor-about-modis-janata-curfew/articleshow/74758687.cms.

Dryzek, John. 2005. "Deliberative Democracy in Divided Societies: Alternatives to Agonism and Analgesia." *Political Theory* 33 (2): 218–42.

Dubash, Navroz. 2012. *Handbook of Climate Change and India: Development, Politics, and Governance*. London: Routledge.

Dubash, Navroz, Radhika Khosla, Ulka Kelkar, and Sharachchandra Lele. 2018. "Evolving Ideas and Increasing Policy Engagement." *Annual Reviews of Environment and Resources* 43:395–424.

"From Democracy to Mobocracy." 2018. *Economic & Political Weekly* 53 (26-27). Accessed July 4, 2021. https://www.epw.in/journal/2018/26-27/editorials/democracy-mobocracy.html.

Gettleman, Jeffrey, Kai Schultz, and Hari Kumar. 2017. "Environmentalists Ask: Is India's Government Making Bad Air Worse?" *New York Times*, December 8, 2017. https://www.nytimes.com/2017/12/08/world/asia/india-pollution-modi.html.

Gokhale, Nihar 2020. "To Kickstart the Economy, India's Environment Ministry Is Clearing Projects in 10 Minutes." *Quartz India*, May 5, 2020. Accessed July 4, 2021. https://qz.com/india/1851634/india-fast-tracks-green-clearance-to-spur-coronavirus-hit-economy/.

Guha, Ramchandra 2000. *The Unquiet Woods: Ecological Change and Peasant Resistance in the Himalaya*. Berkeley: University of California Press.

Gupta, Himangana, Ravinder Kohli, and Amrik Ahluwalia. 2015. "Mapping 'Consistency' in India's Climate Change Position: Dynamics and Dilemmas of Science Diplomacy." *AMBIO* 44:592–99.

Hall, Ian. 2017. "Narendra Modi and India's Normative Power." *International Affairs* 93 (1): 113–31.

Haluza-DeLay, Randolph, Robin Globius Veldman, and Andrew Szasz. 2013. "Social Science, Religions, and Climate Change." In *How the World's Religions Are Responding to Climate Change*, edited by Robin Globius Veldman, Andrew Szasz, and Randolph Haluza-DeLay, 3–19. London: Routledge.

Jaffrelot, Christophe. 2018. "Coalition Country." *Indian Express*, June 23, 2018. Accessed July 4, 2021. https://indianexpress.com/article/opinion/columns/coalition-country-united-opposition-bjp-2019-lok-sabha-elections-5229461/.

Jenkins, Willis, and Christopher Chapple. 2011. "Religion and Environment." *Annual Review of Environment and Resources* 36 (1): 441–63.

Kumar, Chaitanya. 2015. "Modi's Oxymoronic Stance on Climate Change." *Huffington Post*, April 11, 2015. Accessed July 4, 2021. https://www.huffpost.com/entry/modis-oxymoronic-stance-on-climate-change_b_7045276.

Liebig, Michael. 2013. "Kautilya's Relevance for India Today." *India Quarterly* 69 (2): 99–116.

Mazumdar, Arijit. 2018. "India's Soft Power Diplomacy under the Modi Administration: Buddhism, Diaspora, and Yoga." *Asian Affairs* 49 (3): 468–91.

Mehra, Puja. 2015. "G-20 Summit: Modi to Push for Inclusive View on Climate." *The Hindu*.

Miller, Manjari, and Kate de Estrada. 2018. "Continuity and Change in Indian Grand Strategy: The Cases of Nuclear Non-proliferation and Climate Change." *India Review* 17 (1): 33–54.

Mohan, Chilamkuri. 2015. "Modi's Multilateral Moment." *Indian Express*, December 1, 2015. Accessed July 4, 2021. https://indianexpress.com/article/opinion/columns/raja-mandala-narendra-modi-multilateral-moment-paris-climate-talks/.

Mukherji, Nirmalangshu. 2018. "Is the Ghost of Fascism Haunting Political Thought?" *Economic & Political Weekly* 53 (28). Accessed July 4, 2021. https://www.epw.in/journal/2018/28/special-articles/ghost-fascism-haunting-political-thought.html.

Nanda, Meera. 2005. *The Wrongs of the Religious Right: Reflections on Science, Secularism, and Hindutva*. Gurgaon, India: Three Essays Collective.

Narang, Sonali. 2017. "Framing of Migration, Climate Change and Their Implications for India: Rhetoric, Reality and the Politics of Narrative." *Journal of Asia Pacific Studies* 4 (3): 314–59.

Nugteren, Tineke. 2005. *Belief, Bounty, and Beauty: Rituals around Sacred Trees in India*. Leiden, Netherlands: Brill Academic Publishing.

Rahman, Maseeh. 2014. "Indian Prime Minister Claims Genetic Science Existed in Ancient Times." *Guardian*, October 28, 2014, 17.

Saran, Samir. 2010. "Climate's Holy Trinity." *Indian Express*, May 13, 2010. Accessed July 4, 2021. http://archive.indianexpress.com/news/climate-s-holy-trinity/617947/.

Sengupta, Himangana. 2019. "Narendra Modi and India's New Climate Change Norms." Terra Nova, Observer Research Foundation. July 18, 2019. Accessed July 4, 2021. https://www.orfonline.org/expert-speak/narendra-modi-and-indias-new-climate -change-norms-53154/

Sharma, Mukul. 2012. *Green and Saffron: Hindu Nationalism and Indian Environmental Politics*. Ranikhet, India: Permanent Black.

Sivaram, Varun. 2017. "The Global Warming Wild Card." *Scientific American* 316 (5): 48–53.

Sivaramakrishnan, Kalyanakrishnan. 2015. "Ethics of Nature in Indian Environmental History: A Review Article." *Modern Asian Studies* 49 (4): 1261–310.

Sullivan, Kate. 2014. "Exceptionalism in Indian Diplomacy: The Origins of India's Moral Leadership Aspirations." *South Asia* 37 (4): 640–55.

"Terror, by Another Name." 2018. *Economic & Political Weekly*, July 14, 2018. Accessed July 4, 2021. https://www.epw.in/journal/2018/28/editorials/terror-another-name.html.

Venkatasubramanian, Kalpana. 2016. "Examining the Politics of Climate Change: Narratives, Actions and Adaptations in Gujarat, India." PhD diss., Rutgers University.

Vyas, Sharad. 2016. "If You Fight Local Pollution, You Can Effectively Fight Climate Change." *The Hindu*, October 3, 2016. Accessed July 4, 2021. https://www.thehindu.com/sci-tech /energy-and-environment/If-you-fight-local-pollution-you-can-effectively-fight -climate-change-Erik-Solheim/article15088912.ece.

Zimmermann, Francis. 1987. *The Jungle and the Aroma of Meats: An Ecological Theme in Hindu Medicine*. Berkeley: University of California Press

NEERAJ VEDWAN is an environmental anthropologist working in the areas of water resources management, environmental policy, environmental perceptions and attitudes, and impacts of and response to climate change. Dr. Vedwan received his PhD from the University of Georgia, Athens, and subsequently completed a postdoctoral assistantship at the University of Miami, Florida, on the use and dissemination of climate forecasts in agriculture and other areas. Since 2003, he has been working at Montclair State University in the Department of Anthropology as Associate Professor. His research has been published in prestigious journals including *Cultural Anthropology, Human Organization,* and *Water Resources Management.* He is a fellow of the Society for Applied Anthropology and has worked as an external consultant for the European Commission on water contamination issues in India.

II.

TRANSNATIONAL AND THEORETICAL CONSIDERATIONS

4

CAST OUT FEAR

Secularism, (In)security, and the Politics of Climate Change

Erin K. Wilson

The truth is that the most systemic threat to humankind remains climate change and I believe it is my duty to remind it to the whole of the international community.

—Antonio Guterres (United Nations Secretary-General 2018)

The catastrophic scenes from Australia's wildfires should alarm all of us. Climate change is driving even more dangerous and destructive fires across the world, from California to New South Wales—and we must fight together to defeat this crisis.

—Elizabeth Warren, United States Senator and 2020 Presidential Candidate (Doherty 2020)

THREAT. ALARM. DANGER. CRISIS. THESE ARE THE WORDS that characterize the way global political leaders and the media speak about climate change. There is a prevailing sense of "clear and present danger" in high-level conversations about climate change and of the need for urgent action to prevent imminent disaster. We should be concerned about climate change; we should be worried about it. Fear is the overriding emotion.

Where does this fear come from? On one level, it is arguably a logical response to the scientific evidence and prognoses of worst-case scenarios should humanity fail to take collective action, made worse by continued global inertia and political stalemates. Yet, in a sense, this fear is deeper, more fundamental. The fear that is palpable in global political discourses about climate change extends beyond rational, cognitive fear and beyond "threats" to our national security or "risks" that insurance companies protect us against. It does not just emerge from the scientific evidence. It is a fear that is felt and intuitive—somehow embedded in the fabric of our being. It is rooted in the ways that Euro-American cultural and historical perspectives imagine the relationship between humanity and the natural world (Gutkowski 2014).

This relationship in the Euro-American imaginary has long been constructed as binary, separate, at times antagonistic, and almost always unequal. According to this construction, nature is something that humanity should be able to command, control, and subdue. Nature is an object for us to study through science, to harness for industry, to enjoy for leisure (Merchant 1980, 2006). Ultimately, however, it is—or at least should be—subordinated to humanity's will. Nature is wild and unpredictable. Like "women" (Prokhovnik 2003) or "religion" (Wilson 2012), nature can be dangerous and irrational. It must therefore be contained (Merchant 1989, 2003).[1]

For centuries, this view has been predominant. The global economic and political order has been founded on the assumption that humanity is able to control and subdue nature. Climate change represents a potential fundamental disruption to that (perception of) control and the subordination of nature to humanity. That rupture of the relationship between nature and humanity is a crucial part of the reason why we fear climate change (Hulme 2009). It also explains why the proposed responses to climate change revolve around scientific, technological, and economic solutions for adaptation and mitigation. Surely, with our superior knowledge and intellect, humanity will be able to maintain its dominance over nature. Indeed, global climate change leaders such as Christiana Figueres express such sentiments as "our best hope" in meeting the challenge of climate change: "I believe in human ingenuity—that when we decide on a task to be done, no

---

1. This is not the only way of understanding humanity's relationship with nature. Other strains of thought, such as those outlined by Paul Wapner (2010), embrace and celebrate the wildness and unpredictability of nature. Such views are more common among environmentalists, who tend to view the relationship between humanity and nature on more equal terms or view humanity as inferior to nature.

matter how daunting it may seem at the beginning, we are able to unleash human ingenuity and human innovative capacity that was unknown, and takes us to a solution" (cited in McCarthy 2014). Increasingly, however, the realities and consequences of climate change suggest that even our best efforts will not be enough.

Where did this idea come from, that nature and humanity were somehow distinct from one another? That they existed in an oppositional, antagonistic, almost confrontational relationship? Many will (and have) pointed to the influence of the Judeo-Christian religion in Euro-American contexts as a key source for this ontological assumption (Hulme 2009; Mulligan 2012). Christianity, with its command that human beings "be fruitful and increase in number; fill the earth and subdue it. Rule over the fish in the sea and the birds in the sky and every living creature that moves on the ground" (Genesis 1:28, New International Version), is often held responsible for creating this binary between humanity and nature (White 1967, 1205).[2] Despite varied interpretations that suggest this verse is about responsible stewardship rather than dominion, there is still a clear distinction between humanity and nature in the creation stories, one that seems to reinforce humanity's superiority to nature. But Christianity is no longer the overriding "deep culture" (Galtung 1996) in Europe and America that it once was. More and more, the secular has come to dominate these contexts. Despite its own dualistic relationship with religion, arguably, the secular—a space, way of being, and way of thinking in the world that is focused on the immanent and natural, separated from the religious, transcendent, and supernatural—is increasingly understood on its own, without reference to religion, as a normal and natural "self-enclosed reality" (Casanova 2011; Taylor 2007). Has the secular, then, also played a part in maintaining the idea that humanity and nature are somehow distinct and separate, antagonistically opposed to one another?

The prevailing emphasis on "rational"—technological, scientific, economic—policy responses in discussions about how to address climate change provides some support for an affirmative answer to this question. At the same time that "rational," "scientific" responses are privileged, the seemingly intuitive knowledge of environment and place among Indigenous

---

2. It is important to note that this verse, while used as evidence of Christianity's contribution to the subordination of nature, is part of the Hebrew Bible.

peoples and traditional landowners continues to be marginalized (Leonard et al. 2013). This, too, suggests the influence of the secular.

More fundamentally, however, the continued emphasis on climate change as a threat—a danger, a risk, as something to be feared—is also, I would argue, bound up with the dominance of secular ontologies in Euro-American and global political conversations and institutions. Secularism contributes to the production of insecurity and fear precisely because it claims to be able to contain and control what it constructs as irrational, unpredictable, and dangerous. Religion is the primary target; however, through the interrelated binary oppositions on which secular logic rests, nature also becomes something that is to be feared. These binary oppositions that help to create insecurity and fear are also intimately entangled with the emergence of the modern secular nation-state (Mavelli 2011). As such, global political structures that continue to be dominated by the state will inevitably rely on fear (or lack thereof) as a core justification for action (or inaction) on climate change.

This is the argument that I pursue in this chapter. I begin by exploring the relationship between climate change and the politics of fear. Fear is a common emotion and motivator in global politics, crucial to the realist theoretical paradigm and the project of security and securitization. Although not a traditional threat in military or economic terms, climate change and the accompanying alterations it will bring in other domains of global politics are increasingly classified as threats to national security. The politics of fear—fear of resource scarcity, more frequent and intense natural disasters, and increasing displacement and migration—are at the core of this view of climate change as a security threat. However, this view, I suggest, does not just come from a realist "power politics" understanding of global politics and security. It is also embedded in the secular ontologies that have come to dominate global political relationships, institutions, and policy frameworks. I explore these secular ontologies, how they are constituted, and particularly how they relate to the politics of fear and insecurity. This occurs in two key ways: (a) through the binary oppositions on which secular logic rests, particularly the oppositions between culture–nature and reason–emotion, and (b) through the assumed, albeit implicit, positive relationship between secularism and security that is central to the modern state and states system.

This fear and insecurity at the heart of secular ontologies is a crucial part of how we presently understand climate change, not only because we fear climate change but also because we fear nature itself and view nature and humans as existing in a dualistic antagonistic relationship. The primary purpose of the modern secular nation-state is to provide protection from religious violence and the state of nature (Cavanaugh 2009; Mavelli 2011). This assumed positive relationship between secularism and security rests on a fundamentally flawed understanding of religion, one produced by secular ontologies to create the necessity of privatizing and containing religion. Building on these central overlapping binary oppositions of secularism–religion, security–insecurity, humanity–nature, and reason–emotion, the modern state and states system then also contributes to the construction of nature as a threat that can be properly subdued and contained only through state-based politics and governance, privileging "rational" (scientific, technological, and economic) policy responses as the only "proper," "effective" way to respond to the threats posed by climate change.

This analysis points to a fundamental need to rethink each of these relationships, and the various ways that they are entangled, to recognize the limitations of secular ontologies with regard to the relationships of fear, insecurity, and opposition to which it contributes. Until and unless we can conceive of the relationship between humanity and nature in ways that move beyond these binary antagonistic conceptualizations, it will be almost impossible to conceive of alternative responses to climate change beyond the paradigm of security and the politics of fear. In the final section of the chapter, I consider the possible ways forward from this point, neither discarding secular ontologies nor downplaying the very real dangers that are possible if we fail to respond to climate change in any meaningful way. Rather, the goal is to destabilize secular ontologies from their dominant position and create space for other ontologies that understand the relationship between humanity and nature in more equal, harmonious, and cooperative ways. The possibilities that such a reconceptualization may open up how we address climate change are worth serious consideration.

Climate Change and the Politics of Fear

Fear is a common component of the ways in which we think and talk about climate change, both implicit and explicit. Rather than speak of fear, scholars, policy makers, and practitioners will refer to the "challenge" of climate change, the "threat" of climate change, the "risks" it presents, especially if nothing is done to mitigate its impacts. Kate Manzo (2018) highlights the ways in which metaphors drawn from Christian iconography and Greek mythology are utilized in cartoons about climate change to emphasize the sense of threat and danger of "planetary death" from climate change if nothing is done to address it. Climate scientists highlight the plethora of "fear-inducing narratives" that are deployed in the public domain to communicate the risks, dangers, and realities of climate change (O'Neill and Nicholson-Cole 2009). Increasingly, however, scientists are suggesting that this emphasis on fear may be ineffective at best and counterproductive at worst (O'Neill and Nicholson-Cole 2009; see also Norgaard 2011).

This perspective raises a curious question: Why are fear, risk, challenge, and threat the main themes in discussions of climate change, when this may not be the most effective way to communicate about and mobilize people to take action on climate change?

There are, I suggest, two key reasons for the strong presence of fear in narratives about climate change, both of them inextricably tied to secular ontologies. There may well be other reasons, but for now, I shall focus on these two. The first is the central role that fear plays in global politics in general (Bleiker and Hutchison 2008). In military security, economic crises, migration, terrorism, and disease, for example, an underlying sense of fear often drives media discourses and policy making on many of these issues. Arguably, fear may be the most significant emotion in global politics. It is the key motivator and driver of much of our state-based politics, policy making, and action. Oddly, research on emotions in general and fear in particular has, until recently, been largely absent from the study of global politics. This neglect of emotions may largely be attributed to the myth of politics as an arena of rational debate and decision-making, driven by reason rather than emotion. This myth, in turn, stems back to Enlightenment assumptions about rationality and the capacity to separate reason from emotion (Bleiker and Hutchison 2008)—this capability has rarely, if ever, been the case in reality. Nevertheless, the myth of rational, unemotional politics persists.

The goal of state-based politics has primarily been to mitigate and control fear and insecurity. This approach is based on the assumption that perceptions of risk and insecurity are rational. The potential that fear may be unfounded or unreasonable or may be overcome by adopting a different ontological perspective is rarely considered. Consequently, the centrality of fear in the structuring logic of state-based global politics generally is one factor that helps to explain why fear is so central in narratives about climate change.

A second key factor, I suggest, is the way we understand the relationship between humans and nature. In his book *Why We Disagree About Climate Change*, Mike Hulme (2009) explores "the things we fear." He draws on Douglas and Wildavsky's "four 'ways of life'" framework regarding how humans perceive their relationship with their environment and the consequences that has for how we respond to environmental challenges and risks, including climate change. These four ways of life are summarized as "nature is a lottery, capricious; nature is tolerant if treated with care; nature is ephemeral; nature is benign" (Hulme 2009, 186). Hulme then goes on to explain each of the different positions. People who view nature as capricious are described as "fatalists." They see the climate system as fundamentally unpredictable, influenced by many factors, with humans being only one of them. The climate has always presented a risk to humans and will continue to do so. Those who see nature as tolerant are labeled "hierarchists." They argue that greater knowledge about the climate system is needed to effectively manage the risks of climate change. Those who understand nature as ephemeral are labeled "egalitarian." Hulme summarizes their position as viewing the climate system as "precarious and in a delicate state of balance" that can easily be disturbed by humans. Holders of the fourth view, nature as benign, are categorized as "individualists." Overall, they understand any risks or challenges presented by climate change as ultimately manageable (Hulme 2009, 189–90).

At first glance, these four positions represent highly diverse perspectives on the relationship between humans and nature, all with important consequences for how people will respond to the challenges of climate change. Nevertheless, there is one characteristic they all share: by and large, these four views all understand humans and nature as distinct from one another. They may exist in relationship, but they are separated, with the potential for humans to affect nature, to control it, and to respond to it effectively. In this way, humans are implicitly positioned as superior over nature, with the capacity to understand, control, and respond to changes in climate.

This view of the relationship between humans and nature is often referred to as the nature–culture dichotomy and is a central component of Euro-American ontologies (Caillon et al. 2017). In this dichotomy, humans are positioned as superior to nature. Humans are able to control nature, subdue it, and use it for their own needs and purposes. Climate change threatens to disrupt this relationship of superiority and control. There is a risk not only that we might not be able to control climate change but also that we might not be able to effectively adapt to the new climate conditions. It is the potential disruption of this long-established belief in the superiority and capacity of humans to control nature that also contributes to the pervasive sense of fear in narratives about climate change.

This nature–culture dichotomy is also related to the logic of state-based politics. The modern state, particularly within the realist view of world politics, is largely understood as an attempt to control the state of nature. The state brings order to an otherwise anarchical and chaotic system, providing its citizens with security and protection. Here, too, nature becomes something that is to be feared but that the state is able to control and from which it can protect its citizens (Bleiker and Hutchison 2008; Mavelli 2011).

These two dichotomies—reason–emotion and nature–culture—are inextricably linked, in various ways, to the secular ontologies that inform how scholars, policy makers, and citizens in many Euro-American contexts make sense of global politics and, in particular, climate change. The underlying logic of secular ontologies is based on the central dichotomy of secular–religious, of religion as something separate and distinct from the secular. Secular ontologies effectively create religion in order to control and subdue it. That is the core unifying element of all secular ontologies: religion is *something*—a structure, belief system, institution, way of thinking and being—that can be clearly and neatly distinguished and separated from all other realms of human activity. *Religion* understood in this way is effectively a modern Western invention, stemming from the Enlightenment (Asad 2003). In other places and times, religion does not exist. This is not to say that there are no belief systems or ways of being in the world that are explicitly conscious of divine beings, supernatural forces, and transcendence. Rather, these ways of being are entangled with all other realms of human activity, not separated out and divided from them. Having established religion as a separate and distinct category, secular ontologies then subordinate religion to the secular. This subordinating move is achieved by linking religion with the inferior half of other dichotomies—nature, emotion, tradition, women—and the secular with the superior half—culture, reason, progress, men (Wilson 2012, 2017). And yet, ultimately, secularism fails to establish complete control, in part because the secular itself is derived from the religious. Certain religious (Christian) leaders, such as Pope

Francis, and religious (Christian) narratives retain positions of power and influence in climate politics, if only in symbolic ways. In the next section, I elaborate on what I mean by "secular ontologies" and the ways in which these binary oppositions contribute to the production of fear in discourses on and responses to climate change.

Secular Ontologies, Climate Change, and Insecurity

Secularism has arguably been the dominant model for liberal statecraft and a powerful ideology structuring Euro-American political communities since the Enlightenment. Secularism as statecraft is concerned with laws and institutions that manage relationships between religious and secular authorities and domains (Casanova 2011). Although the two do not necessarily overlap in theory, in practice, secular statecraft is frequently underpinned by variations of what Jose Casanova (2011) calls "secular ideologies."

Multiple attempts have been made to describe these overarching framing assumptions, narratives, and practices that are often referred to as the *secular*—as worldviews, ideologies, or philosophies. Secularism, however, is more than an ideology. It is, I argue, an all-encompassing ontology. Understanding secularism as an ontology enables us to see beyond the normative assumptions about the value of the "religious" and the "secular." It allows us to see how these assumptions are enacted not only through the ways in which people think and understand the world but in how they behave, occupy spaces, and embody practices, particularly in their interactions with others and with nature. In addition, describing secularism as an ideology, philosophy, or worldview carries with it ideas of transience and changeability. Ideologies and philosophies are usually consciously held and articulated belief structures. An ontology, by contrast, is frequently not articulated or consciously held. Ontologies are more intuitive, ingrained in the fiber of a person's being largely as a result of being brought up and living in their midst. There is something about an ontology that one feels rather than knows, or perhaps knows through feeling or experience.

Ontology seems to be one of the few concepts that is able to encompass both secular and religious ways of understanding the world. It offers us a way to think about these terms as different but interconnected components of a broader category rather than as diametrically opposed ways of understanding the world. Understanding secularism as an ontology, then, offers a way to address the often implicit conceptual and normative inequality between the secular and the religious. Religion and other categories associated with it (including emotion and nature) are predominantly subordinated to the secular and its associated categories (Wilson 2012, 2017). This is not to say that ideology and worldview are not useful ways of describing secularism or that ideological secularism does not exist. What I am attempting to describe by using ontology transcends ideology.

Since the early 2000s, scholars in religious studies, philosophy, and international relations have argued that no version of secularism provides a neutral, universal basis for public reason, contrary to long held liberal assumptions (Casanova 2011; Connolly 1999; Eberle 2002; Hurd 2008; Kuru 2007; Taylor 2009; Wilson 2012). Rather, secularism represents "fundamental shifts in conceptions of self, time, space, ethics, and morality" (Mahmood 2016, 3). Secularism is a highly specific, culturally embedded model for managing the relationship between religion and politics that emerged in Euro-American contexts as part of the Enlightenment but that has become influential across diverse regions of the world (Gutkowski 2014, 6). In other words, secularism is a distinctive ontology, or theory about what exists (Pedersen 2001, 413). It "redefines and transcends particular and differentiating practices of the self that are articulated through class, gender, and religion" (Asad 2003, 21–22); it constitutes particular practices and ideas along the natural–supernatural binary, positioning some practices within the category of the natural or the secular, whereas others are placed in the category of the supernatural—religion, superstition, or fetishism.

Secular ontologies attribute particular characteristics to these practices—irrational, violent, chaotic, divisive (Wilson 2012). These inherent assumptions have come to dominate how we analyze practices constructed as "religious" and how they intersect with and affect politics and public life. Secularism's origins within the Euro-American context contribute to its association with colonialism and with binary oppositions between not only the secular and the religious but also modern–primitive, reason–emotion, Western–non-Western, and nature–culture (Wilson 2012) that continue to affect power relations in global politics. These binary oppositions are also a crucial component of the logics that govern our conceptualizations of and responses to the phenomenon labeled *climate change*.

This is not to suggest that secularism is monolithic, homogeneous, or exclusively Western. Like religion, secularism is not a singular entity. Secular ontologies are diverse, shifting, changing, unstable, and contextually specific (Daulatzai 2004, 567). Indeed, although secularism emerged from local contexts and historical trajectories in Europe and the United States, through globalization, it has merged to constitute a globalized agglomeration of ideas and practices that vary locally. What secularism means in the Netherlands, for example, is different from what it means in India, Bangladesh, France, Canada, and so on (Hurd 2008; Kuru 2007).

At the same time, although secularism does not mean the same thing from one place to the next, certain "family resemblances" characterize secular ontologies across their different manifestations:

1. Religion is something tangible and identifiable, that can be clearly distinguished, defined, and separated from the secular, which can also be clearly defined.
2. In addition, religion should be clearly distinguished and separated from other areas of human activity, such as politics, economics, law, education, and so forth, that are grouped under the secular (Asad 2002, 116).
3. This is because religion is subjective, particular, individual, and irrational (Hurd 2008; Wilson 2012), as opposed to the secular, which is neutral and universal.
4. Also, religion is what people disagree about more frequently and violently than anything else (Cavanaugh 2009), thus religion is the fundamental cause of violence, intolerance, and chaos.
5. Religion must therefore be kept out of the public sphere and relegated to the private to preserve order and peace (Taylor 2009; Wilson 2012), meaning that the distinction between religion and the secular is managed through the existence of public and private spheres (that are equally as unstable and problematic as categories of religion and the secular).
6. Finally, religion is always subordinated to the secular in that, even if religion is viewed as something that can contribute positively to politics and public life, religion's interventions should still be regulated by so-called secular authorities and institutions.

Secular ontologies have also contributed to the production of hierarchies of religions, so certain religions are constructed as more threatening and therefore more to be feared, whereas others are given positions of power and influence in climate politics, even if only symbolically. As Corey Robin (2004) has highlighted, hierarchical power structures and inequalities are crucial components of the production and maintenance of political fear. The secular and secular ontologies emerged from Christian ontologies (Casanova 2011). As such, despite every effort to separate the secular from the religious, it is almost impossible for the secular to escape religion. The origins of secular ontologies within Christian ontologies also help to explain why Christian voices and narratives are often privileged in climate politics, whereas other religious voices, particularly Muslim and Indigenous, are marginalized, ignored or actively belittled or reviled. Nevertheless, the dependency that secular ontologies have on religious or Christian ontologies is rarely, if ever, acknowledged.

Binary logic regarding the relationship between religion and the secular is a core element in the underlying assumptions of secular ontologies. It is through this binary logic that secular ontologies become implicated in the ways in which we understand and respond to climate change. Indeed, secular ontologies are arguably at the heart of fear- and security-driven responses to climate change that dominate much of current political and social discourses and policy.

To date, prevailing proposals for how to respond to climate change have largely emphasized scientific, technocratic, and economic strategies (Steger, Goodman, and Wilson 2013). These strategies include addressing historical injustices in carbon emissions through technology transfers from global North to global South to compensate for these inequalities and assist with mitigation and adaptation to the effects of climate change. This emphasis on science, technology, and the economy stems in no small measure from the secular ontological assumptions I have outlined. These assumptions, which are also connected with other frameworks such as modernism and colonialism (Asad 2003; Wilson 2012), have come to dominate particular arenas of global politics and policy, including climate change (Ager and Ager 2011, 2015; Gutkowski 2014). They shape who is considered to be a "legitimate" actor in global politics, what kinds of knowledge and evidence are viewed as acceptable, and thus, crucially, the kinds of policy responses that are deemed appropriate. The overlap between the binary oppositions of secular–religious, reason–emotion, and culture–nature, along with male–female and traditional–modern, all contribute to the privileging of particular actors and sources of knowledge and the exclusion and marginalization of other actors and kinds of knowledge.

First, secular ontologies affect who we deem to be religious or nonreligious actors. This designation has multiple implications in the area of human rights law, especially freedom of religion or belief (Sullivan 2005; Hurd 2015), but also affects climate change politics. Initiatives to "engage religious actors" in action on climate change make assumptions about who these religious actors are, which can lead to the exclusion and marginalization of other actors who are not "religious" but also do not fit the accepted mold of a rational, secular political actor. Once this distinction between religious and nonreligious actors is made, religious actors are then imbued with particular characteristics on the basis of the core secular assumptions. These can range from traditional secular assumptions that religious actors are "irrational, violent, conservative, patriarchal" or more recent assumptions stemming from the "good religion/bad religion" view (Fiddian-Qasmiyeh 2014; Hurd 2015), in which religious actors are deemed good or bad depending on how well they align with prevailing secular liberal norms and values. These assumed characteristics affect the position an actor has and how they are viewed and valued within global political arenas.

Secular ontologies also influence the kinds of knowledge and evidence that are considered acceptable or reliable. Knowledge that stems from science and scientific research is deemed more reliable in some contexts than forms of knowledge drawn from local cosmologies and religious ontologies (although here, too, there is a danger of romanticizing local Indigenous cosmologies, particularly within the climate justice movement). This approach influences practical programs, funding, collaboration, and which actors are given a platform and which are not. Certain religions, in particular Christianity, and religious actors enjoy a level of acceptance and legitimacy within secular ontologies and institutions influenced by secular ontologies because they are interconnected and the one arguably grew out of the other (Asad 2003; Casanova 2011). It is also important to highlight that religious actors themselves can reinforce the assumptions of secular ontologies by, for example, endeavoring to demonstrate their "added value" to secular programming. Other religious actors do not enjoy this same level of acceptance and legitimacy. Consequently, inequalities exist not only between secular and nonsecular ontologies but also among different religions within the dominant secular frame.

Secular ontologies can contribute to erroneous understandings of who the trusted and legitimate actors are in particular local contexts—an issue that is particularly significant for efforts to promote climate justice (Daulatzai 2004). For actors and institutions embedded in secular ontologies, so-called secular actors—nongovernmental organizations, grassroots movements, civil society networks, universities, and, to a lesser extent, states—are the trusted power brokers, negotiators, and mediators. This relationship can lead to the marginalization of religious actors or actors who do not conform to the strict secular–religious binary in research, policy, and advocacy by institutions and actors embedded within secular ways of being. However, this is not the case the world over. In many small island nations (Nunn 2017) and in countries in the Middle East (Sadiki 2009) and on the African continent (Bartelink and Wilson 2020), states and secular actors are often distrusted, in part because of a history of failed secular politics. In addition, in these contexts, secular ontologies often just don't make sense. They are not the way people understand the world. Many people do not inhabit secular worlds.

This observation indicates another way in which secular ontologies are implicated in the politics of climate change and, in particular, with reference to fear and insecurity. Secularism is entangled with the modern state and states system that is so central to contemporary global politics. The standard narrative within international relations regarding the rise of the modern state is closely tied to narratives about the rise of secularism and the decline of religion. The so-called "wars of religion" in the sixteenth and seventeenth centuries are a pivotal moment in this narrative, culminating in the Peace of Westphalia of 1648. Westphalia, so the story goes, was the point at which religion, for the most part, was contained and controlled by the secular state. The wars of religion proved definitively that religion in the public realm leads to violence, intolerance, and chaos, and thus the secular state is necessary to maintain peace and public order (Hurd 2008; Mavelli 2011; Wilson 2012).

This narrative has multiple problems, as numerous scholars have highlighted. The wars of religion were not really—or, at least, not only—about religion but about trade, political alliances, power, and territory (Asch 1997; Thomas 2000). The Peace of Westphalia, arguably, was not the definitive moment in the rise of the modern state or the privatization of religion because elements of the modern state are visible well before 1648, and characteristics of feudal and imperial regimes are evident well after (Krasner 1993; Philpott 2001; Wight 1977). Nevertheless, the narrative remains powerful.

Central to this narrative is the implication that secularism ensured security, in that it privatized religion and thereby provided protection from the violence and chaos that religion generates. Secularism thus gained dominance over religion. As Luca Mavelli (2011) writes, "By establishing a connection between religion in its public and political manifestations and violence, these perspectives implicitly posit a positive relationship between security and secularization" (178). This positive relationship between secularism and security rests on a fundamentally flawed understanding of religion and of European history and the emergence of the modern nation-state

(Asad 2003; Mavelli 2011, 179). Rather than resolving the problem of religious violence, this understanding of the relationship between secularism and security in some respects generates fear and insecurity. This is not accidental. The raison d'etre of the modern nation-state is to provide security for its citizens. This role of providing protection and security from the things we fear is what endows the state with power and legitimacy in contemporary global politics. In recognizing that contemporary accounts of security and the secular state rest on problematic and flawed understandings of religion, we can see that secular ontologies also contribute to the production of fear and insecurity. Secularism creates the very situation from which people must be protected (Mavelli 2011). As Talal Asad (2003) observes, "The difficulty of secularism as a doctrine of war and peace in the world [and thus arguably a doctrine of security] is not that it is European (and therefore alien to the non-West), but that it is closely connected with the rise of a system of capitalist nation-states—mutually suspicious and grossly unequal in power and prosperity, each possessing a collective personality that is differently mediated and therefore differently guaranteed and threatened" (9).

Binary oppositions are again central to this connection between secular ontologies, the modern state and security. As noted, secular ontologies create "religion" as something tangible, real, distinguishable, and identifiable. They then endow "religion" with particular characteristics that justify religion's subordination to the secular. In doing so, secular ontologies work in collaboration with other binary logics of exclusion, such as modernism, colonialism, racism, rationalism, and patriarchy to subordinate sociopolitical actors and elements deemed to be religious, premodern, non-Western, nonwhite, irrational, emotional, natural, and feminine. This subordination occurs through constructing these actors and elements as threats to the peace, security, and stability of the secular, modern, rational, cultural, masculine, public, political domain. It is no coincidence that nature is often described as feminine, with wild, unpredictable, and emotional characteristics. Nature becomes one more thing in the list that we should fear and from which the modern state is responsible for protecting us.

What this analysis suggests is that not only are secular ontologies a crucial component of how we understand "religion" in contemporary politics, but also they are fundamentally entangled in the logic of the modern state, security, and the politics of fear. Through this dominance of security, along with the opposition between humans and nature, secular ontologies are a central structuring logic in global discourse and policy making around how we understand and respond to climate change. While secular ontologies remain dominant, largely unrecognized and uncontested, it will be difficult, if not impossible, to conceive of climate change as anything other than a threat we should fear and from which the state must protect us. This view will make it equally difficult to move beyond scientific, technological, and economic fixes to the problems presented by climate change—fixes that are already proving to be inadequate in the face of the scale and speed of the changes that are occurring. Exploring alternative ways of conceiving the relationship between humans and nature beyond secular ontologies becomes crucial in moving past the politics of fear and insecurity.

Multiple Ontologies and Alternatives

I have suggested thus far that secular ontologies play a critical role in structuring how we think about the problem of a changing climate. Secular ontologies are, in some ways, responsible for the binary way in which the relationship between humans and nature is imagined. Furthermore, secular ontologies contribute to the sense of threat and fear with which we view the prospect of a changing climate because of the power relationships they construct between humanity and nature and the central role of the secular state in controlling and containing the sources of insecurity. Secular ontologies further contribute to the exclusion of other nonsecular ontologies by constructing them as irrational, emotional, and premodern, thereby marginalizing and excluding alternative ways of being and knowing in relation to a changing natural world. To address the politics of fear that is currently so central in discourses and policy making around climate change, alternative ontologies should be explored to conceive of the relationship between humans and nature in different, more equal, and less fearful ways.

It is important to stress that, like other scholars who are critical of secularism, I am not suggesting that secularism should be dispensed with, nor am I unconscious of the many important achievements that secularism has enabled. Secular approaches to public life are bound up with questions of justice and equality—how to protect freedom of conscience, how to equally value all religions (as problematic as that goal may be; Sullivan 2005). As Saba Mahmood (2016) notes, "To critique a particular normative regime is not to reject or condemn it; rather, by analyzing its regulatory and productive dimensions, one only deprives it of innocence and neutrality so as to craft, perhaps, a different future" (21). Critiques of secularism are an attempt to recognize the vulnerabilities and shortcomings of secularism, so as to contribute to the development of alternative, more inclusive futures. It is precisely these more inclusive futures that are desperately needed in our sociopolitical imaginations on the issue of climate change.

Three core assumptions of secular ontologies need to be dismantled in relation to the particular challenge of climate change. The first is the binary understanding of humans and nature as separate and that humans can and should control nature for their own purposes. Scientists are increasingly recognizing this understanding of the human relationship with nature as a barrier to effective communication and collaboration to address climate change across different cultures (Caillon et al. 2017; O'Neill and Nicholson-Cole 2009). Furthermore, understanding the relationship between humans and nature in this way may be severely limiting with regard to the possibilities for how we can respond to climate change (Caillon et al. 2017; Leonard et al. 2013). Climate scientists are turning to anthropology, specifically to in-depth ethnographic accounts of nonsecular ontologies that conceptualize relations between humans and nature as inextricable, even seeing them as one rather than as separate. Engaging these alternative ontologies enables us to broaden the scope of how we understand the relationship between humans and nature (Caillon et al. 2017; Hulme 2009).

The second assumption concerns the distinctions between reason and emotion, secular and religious, and modern and traditional. These overlapping binaries generate the view that modern rational (masculine) science, technology, and economy are the most effective (potentially, even the only) options for dealing with climate change. The third assumption relates to the centrality of the state in secular ontologies—that the state is the provider of protection from threats from the potentially dangerous elements in all these binaries (nature, religion, emotion, tradition, and women).

Placing these three core assumptions of secular ontologies in question and recognizing their potential limitations is a first step toward creating space for alternative perspectives of how we conceptualize and respond to the changes occurring in nature linked to a changing climate. Anthropologists working among Indigenous peoples have long observed that the ways in which people understand their relationship with land, animals, plants, humans, and spirits is completely different from dominant, modern, Western, secular structures and categories for understanding the world (Caillon et al. 2017, 26). Many of the Indigenous clans and communities that inhabit parts of the Australian continent view the connection between people and land as continuous and holistic, not as separate and distinct. The well-being of the land is directly related to the mental and physical health of the people (Petheram et al. 2010). Elizabeth Povinelli (1995) has written on the dynamic ways in which Indigenous Australians relate to land, in which the land and the spirits and ancestors that inhabit the land speak to those who are alive in the present (see also Grant 2016). Indigenous ontologies do not fear the land and nature but rather revere and honor them, seeing land and nature not just as an important backdrop to their existence but as entangled with social and communal identities. Likewise, however, Povinelli (2001) has highlighted the limitations of government and science to take the perspectives of Indigenous peoples seriously; ultimately, the logic of science and the economy are given greater weight, reinforcing the inequalities between different types of knowledge that are engendered by the binary oppositions at the core of secular ontologies. Other Indigenous ontologies understand particular beings as people that Western ontologies would label as "animals." Descola (1986/1994), for example, writes about the Achuar in South America, who hold that animals have the same spirit (*wakan*) as humans and appear as animals only on the outside. Numerous other examples are also influencing the development of policy and environmental law, where attempts are made to "give nature a voice" in decision-making processes (Caillon et al. 2017) by acknowledging and taking seriously the worlds inhabited by Indigenous peoples—worlds that are wholly other from the world of secular ontologies (Wilson 2017).

What I am proposing, following Mario Blaser's (2013) work on political ontology, is the application of a multiple-ontologies approach in the realm of climate change science, discourse, and policy responses as a means to address the dominance of secular ontologies and the associated problems that this dominance presents. A multiple-ontologies approach represents what Blaser (2013) has termed a "politicoconceptual problem" (548). It requires holding two seemingly contradictory commitments at the same time—holding fast to the assumptions and commitments of one's own ontology while recognizing that the assumptions and commitments of other—perhaps wholly contradictory—ontologies and worlds are equally as valuable; that our own ontology may not be the "correct" or "real" version of reality. This approach necessitates suspending the modern, secular, rational, scientific assumption that there is "one world" out there that can be known and discovered through objective, neutral scientific methods and theorizing. A multiple-ontologies approach requires a commitment from participants to take one another's worlds seriously and value them equally. It requires a willingness to acknowledge that no single ontology or way of experiencing the world holds all the answers to the present crisis that we face. It requires a move away from Cartesian dualistic "either/or" thinking to "both–and" (Wilson 2012). It is not that different ontologies sit in opposition to one another. It is not that, in some situations, a secular scientific rationalist ontology should be applied whereas in others an Indigenous, religious, spiritual, or other nonsecular way of worlding should be applied. Rather, it is drawing on each different ontology, in conversation and collaboration with one another, to develop workable responses to the various climate-related challenges that exist, including the process of identifying what those challenges are to begin with and then how to prioritize them. What these collaborations look like will differ across different contexts. Although climate change is a global phenomenon, its consequences and effects play out differently in diverse locations. Equally, though, the ontologies that are significant in each location will also differ.

Multiple ontologies also require that we do not attempt to relate different ontologies to one another through language and concepts that belong to only one of them. It is not enough to simply explain the science of climate change to peoples and cultures whose ontologies differ using their own concepts and language. Scientists, scholars, media, and policy makers must also understand other ontological perspectives on the changes that are being observed in the natural world, which may or may not be understood as climate change.

It is important to stress that I am not making an argument for cultural relativism. Cultural relativism forecloses any attempt at cross-ontological communication by assuming "that's just what they believe." In many respects, cultural relativism compounds the exclusion of other ontological perspectives on climate change. By simply seeing two ontologies as different and irreconcilable in some respects, cultural relativism implies that an ontology different from one's own is not worth engaging with and trying to understand. It also does nothing to destabilize secular ontologies as the central ways in which climate change is presently understood and addressed. Consequently, cultural relativism does not address the dualistic understanding of the relationship between humans and nature nor the dominance of fear in climate change discourses. Rather, a multiple-ontologies approach requires taking all ontologies seriously and entertaining the prospect that the world may be different from how we understand and inhabit it—that it may, in fact, be an entirely different world.

This does not mean abandoning our own ontological position. I am not at all suggesting that we should do away with science or secularism. What I am suggesting, however, is that we recognize their limitations and acknowledge and learn from other ontologies, beginning with altering the way in which we understand the relationship between humans and nature. Secular worlds are very different from, for want of a better term, *nonsecular* worlds. If we are genuinely committed to climate justice and addressing the moral and political challenges that climate change brings, then it is important that we take these nonsecular worlds and their understandings and explanations of nature and the changing climate seriously on their own terms. We do not necessarily have to agree with them, and we should not fall back into cultural relativism when we do disagree with them. But taking different ontologies and their associated beliefs and practices about the world seriously is a crucial component of developing meaningful communication and shared, just responses on climate change.

In addition, understanding the ontologies that operate in different contexts is essential for identifying the power brokers, the translators, the mediators, and the main issues that people are really concerned about and how climate and climate change and the relationship between humans and nature are understood and imagined (or not) in different contexts. For Indigenous Australians in Arnhem land, for example, issues of poverty and historical mistreatment of peoples and land are as important, if not more so, as contemporary challenges related to climate change (Petheram et al. 2010). For them, effectively addressing climate change must be done in relation to these other problems. This perspective highlights another potential shortcoming of secular ontologies with regard to climate change—the emphasis on scientific, technological, and economic fixes does little to address other concerns such as historical and intergenerational injustices that are bound up with how the effects of climate change will be experienced differently and more acutely by those who are poor and marginalized. Acknowledging the limitations of secular ontologies and taking seriously the accounts that emerge from other ontologies enables us to more effectively consider how we can address the changes brought by climate change but also how we can do so in just and equitable ways for all peoples.

Once commitment to these foundational ways of working is established, collaborators and partners can begin discerning a common language and mode of understanding across their different worlds. This process could begin by identifying the key concerns of different groups but also why these concerns exist. Understanding the "why" provides insight into the underlying ontological positions and views. Through these preliminary interactions, communities can develop shared analysis of the challenges that incorporate insights from the different ontologies. Once this analysis is established, the communities can begin to develop collective responses that, again, include contributions from different ontologies. How these collaborations work in practice will depend on the contexts in which they operate, but this preliminary framework for multiple ontologies provides a means for beginning this work in practice.

Conclusion

As many have pointed out, climate change cannot adequately be addressed by relying solely on scientific, economic, technological, or even political responses. It requires moral responses, but it also requires creative and imaginative responses—responses that reconfigure our understanding of the relationship between humanity and the natural world, that replace fear and insecurity as the motivating forces for action with hope and a shared sense of empathy and community across different human and natural worlds. Climate change raises questions about our capacity to imagine different shared futures together and to ensure that individuals, communities, and nature have the space and the right to imagine those futures together. How we imagine those futures and how we ensure that there is space for all— religious, secular, rich, poor, Black, White, male, female, and all the other categories people use from within their own ontologies to make sense of the world—to engage in that imaginative process is one of the most vital challenges for the global community.

I have suggested in this chapter that secular ontologies are a crucial component of the way in which we presently make sense of and respond to climate change. Secular ontologies contribute to conceptualizing the relationship between humans and nature as separate, distinct, and often antagonistic, which in turn contributes to fear, threat, challenge, and risk as the main ways in which we understand climate change. In addition, secular ontologies rely on binary distinctions between reason and emotion and modern and traditional to reinforce the central binary of secular and religious. Reason, modern, the secular, and humans (especially men) are privileged above emotion, tradition, religion, and nature (and women). This logic helps to exclude particular kinds of knowledge and evidence as illegitimate from contemporary politics and thus as inappropriate for responding to climate change. Finally, secular ontologies are inextricably tied to the foundational logic of the modern secular nation-state, which rests primarily on a logic of fear and insecurity. Contrary to dominant views in international relations and global politics that secularism provides security from the chaos of religious violence and intolerance, secular ontologies contribute to creating fear and insecurity in the very ways in which they construct the problem of "religion" and its related characteristics in the first place.

Secular ontologies also generate hierarchies of religions, privileging Christianity and marginalizing other religions, especially Islam and indigenous ontologies. This leads to certain Christian voices and narratives having (symbolic) influence in climate politics while other non-Christian voices are ignored. Addressing the dominance of secular ontologies and their associated assumptions about nature, reason, and security opens up space for alternative ontologies to contribute to rethinking the relationship between humanity and nature, to move beyond the politics of fear and insecurity and engage in collaborative egalitarian processes for imagining different futures that incorporate the realities of a changing climate and natural world.

References

Ager, Alastair, and Ager, Joey. 2011. "Faith and the Discourse of Secular Humanitarianism." *Journal of Refugee Studies* 24 (3): 456–72. doi:10.1093/jrs/fer030.

———. 2015. *Faith, Secularism, and Humanitarian Engagement: Finding the Place of Religion in the Support of Displaced Communities*. Basingstoke, UK: Palgrave.

Asad, Talal. 2002. "The Construction of Religion as an Anthropological Category." In *A Reader in the Anthropology of Religion*, edited by M. Lambek, 114–32. London: Blackwell.

———. 2003. *Formations of the Secular: Christianity, Islam, Modernity*. Stanford, CA: Stanford University Press.

Asch, Ronald. 1997. *The Thirty Years' War: The Holy Roman Empire and Europe 1618–1648*. London: Macmillan.

Bartelink, Brenda, and Erin K. Wilson. 2020. "The Spiritual Is Political: Reflecting on Religion, Gender and Secularism in International Development." In *International Development and Local Faith Actors: Ideological and Cultural Encounters*, edited by Kathryn Kraft and Olivia J. Wilkinson, 45–58. London: Routledge.

Blaser, Mario. 2013. "Ontological Conflicts and the Stories of Peoples in Spite of Europe: Towards a Conversation on Political Ontology." *Cultural Anthropology* 54 (5): 547–68.

Bleiker, Roland, and Emma Hutchison. 2008. "Fear No More: Emotions and World Politics." *Review of International Studies* 34 (1): 115–35.

Caillon, Sophie, Georgina Cullman, Bas Verschuuren, and Eleanor J. Sterling. 2017. "Moving beyond the Human–Nature Dichotomy through Biocultural Approaches: Including Ecological Wellbeing in Resilience Indicators." *Ecology and Society* 22 (4). https://doi .org/10.5751/ ES-09746-220427.

Casanova, Jose. 2011. "The Secular, Secularizations and Secularisms." In *Rethinking Secularism*, edited by C. Calhoun, M. Juergensmeyer, and J. Vanantwerpen, 54–74. Oxford: Oxford University Press.

Cavanaugh, William T. 2009. *The Myth of Religious Violence*. Oxford: Oxford University Press.

Connolly, William E. 1999. *Why I Am Not a Secularist*. Minneapolis: University of Minnesota Press.

Daulatzai, Anila. 2004. "A Leap of Faith: Thoughts on Secularistic Practices and Progressive Politics." *International Social Science Journal* 182:565–76.

Descola, Philippe. (1986) 1994. *In the Society of Nature: A Native Ecology in Amazonia*. Translated by Nora Scott. Cambridge, UK: Cambridge University Press.

Doherty, Ben. 2020. "From Tina Arena to Elizabeth Warren: The Big Names Weighing in on Australia's Bushfire Crisis." *Guardian*, January 4, 2020. Accessed January 17, 2020. https://www.theguardian.com/australia-news/2020/jan/04/from-tina-arena-to -elizabeth-warren-the-big-names-weighing-in-on-australias-bushfire-crisis.

Eberle, Christopher J. 2002. *Religious Conviction in Liberal Politics*. Cambridge, UK: Cambridge University Press.

Fiddian-Qasmiyeh, Elena. 2014. *The Ideal Refugees: Gender, Islam, and the Sahrawi Politics of Survival*. Syracuse, NY: Syracuse University Press.

Galtung, Johan. 1996. *Peace by Peaceful Means: Peace and Conflict, Development and Civilization*. London: Sage.

Grant, Stan. 2016. *Talking to My Country*. Melbourne: Harper Collins.

Gutkowski, Stacey. 2014. *Secular War: Myths of Religion, Politics, and Violence*. London: I. B. Tauris.

Hulme, Mike. 2009. *Why We Disagree about Climate Change*. Cambridge, UK: Cambridge University Press.

Hurd, Elizabeth S. 2008. *The Politics of Secularism in International Relations*. Princeton, NJ: Princeton University Press.

———. 2015. *Beyond Religious Freedom: The New Global Politics of Religion*. Princeton, NJ: Princeton University Press.

Krasner, Stephen D. 1993. "Westphalia and All That." In *Ideas and Foreign Policy: Beliefs, Institutions, and Political Change*, edited by Judith Goldstein and Robert O. Keohane, 235–64. Ithaca, NY: Cornell University Press.

Kuru, Ahmet T. 2007. "Passive and Assertive Secularism: Historical Conditions, Ideological Struggles, and State Policies toward Religion." *World Politics* 59 (4): 568–94.

Leonard, Sonia, Meg Parsons, Knut Olawsky, and Frances Kofod. 2013. "The Role of Culture and Traditional Knowledge in Climate Change Adaptation: Insights from East Kimberley, Australia." *Global Environmental Change* 23 (3): 623–32.

Mahmood, Saba. 2016. *Religious Freedom in a Secular Age: A Minority Report*. Princeton, NJ: Princeton University Press.

Manzo, Kate. 2018. "Climate." In *Visual Global Politics*, edited by Roland Bleiker, 55–61. London: Routledge.

Mavelli, Luca. 2011. "Security and Secularization in International Relations." *European Journal of International Relations* 18 (1): 177–99.

McCarthy, Shawn. 2014. "Christiana Figueres: Passionate—and Impatient—about Climate." *Globe and Mail*, October 10, 2014. Accessed September 3, 2018. https://www .theglobeandmail.com/report-on-business/careers/careers-leadership/christiana -figueres-passionate-and-impatient-about-climate/article21066366/.

Merchant, Carolyn. 1980. *The Death of Nature: Women, Ecology, and Scientific Revolution*. New York: Harper and Row.

———. 1989. *Ecological Revolutions: Nature, Gender, and Science in New England*. Chapel Hill: University of North Carolina Press.

———. 2003. *Reinventing Eden: The Fate of Nature in Western Culture*. New York: Routledge.

———. 2006. "The Scientific Revolution and *The Death of Nature*." *Isis* 97:513–33.

Mulligan, Martin. 2012. "An Affective Approach to Climate Change." *Local-Global* 10:10–16.

Norgaard, Kari. 2011. *Living in Denial: Climate Change, Emotions, and Everyday Life*. Boston: MIT Press.

Nunn, Patrick D. 2017. "Sidelining God: Why Secular Climate Projects in the Pacific Islands are Failing." *The Conversation*, May 16, 2017. Accessed June 22, 2021. https://theconversation.com/sidelining-god-why-secular-climate-projects-in-the -pacific-islands-are-failing-77623.

O'Neill, Saffron, and Sophie Nicholson-Cole. 2009. "'Fear Won't Do It': Promoting Positive Engagement with Climate Change through Visual and Iconic Representations." *Science Communication* 30 (3): 355–79.

Pedersen, Morten A. 2001. "Totemism, Animism and North Asian Indigenous Ontologies." *Journal of the Royal Anthropological Institute* 7 (3): 411–27.

Petheram, Lisa, Kerstin K. Zander, Bruce M. Campbell, Chris High, and Natasha Stacey. 2010. "'Strange Changes': Indigenous Perspectives of Climate Change and Adaptation in NE Arnhem Land (Australia)." *Global Environmental Change* 20 (4): 681–92.

Philpott, Daniel. 2001. *Revolutions in Sovereignty: How Ideas Shaped Modern International Relations*. Princeton, NJ: Princeton University Press.

Povinelli, Elizabeth A. 1995. "Do Rocks Listen? The Cultural Politics of Apprehending Australian Aboriginal Labor." *American Anthropologist* 97 (3): 505–18.

———. 2001. "Radical Worlds: The Anthropology of Incommensurability and Inconceivability." *Annual Review of Anthropology* 30:319–34.

Prokhovnik, Raia. 2003. *Rational Woman: A Feminist Critique of Dichotomy*. Manchester, UK: Manchester University Press.

Robin, Corey. 2004. *Fear: The History of a Political Idea*. Oxford: Oxford University Press.

Sadiki, Larbi. 2009. *Rethinking Arab Democratization: Elections without Democracy*. Oxford: Oxford University Press.

Steger, Manfred B., James Goodman, and Erin K. Wilson. 2013. *Justice Globalism: Ideology, Crises, Policy*. New York: Sage.

Sullivan, Winnifred Fallers. 2005. *The Impossibility of Religious Freedom*. Princeton, NJ: Princeton University Press.

Taylor, Charles. 2007. *A Secular Age*. Cambridge, MA: Belknap.

———. 2009. "The Polysemy of the Secular." *Social Research* 76 (4): 1143–66.

Thomas, Scott M. 2000. "Taking Religion and Cultural Pluralism Seriously: The Global Resurgence of Religion and the Transformation of International Society." *Millennium: Journal of International Studies* 29 (3): 815–41.

United Nations Secretary-General. 2018. "Secretary-General's Press Encounter on Climate Change." United Nations, March 29, 2018. https://www.un.org/sg/en/content/sg/press-encounter/2018-03-29/secretary-generals-press-encounter-climate-change-qa.

Wapner, Paul. 2010. *Living through the End of Nature: The Future of American Environmentalism*. Cambridge, MA: MIT Press.

White, Lynn. 1967. "The Historical Roots of Our Ecological Crisis." *Science* 155 (3767): 1203–7.

Wight, Martin. 1977. *Systems of States: Edited and with an Introduction by Hedley Bull*. Leicester, UK: Leicester University Press for the London School of Economics and Political Science.

Wilson, Erin K. 2012. *After Secularism: Rethinking Religion in Global Politics*. London: Palgrave.

———. 2017. "'Power Differences and the Power of Difference': The Dominance of Secularism as Ontological Injustice." *Globalizations* 14 (7): 1076–93.

ERIN K. WILSON is Associate Professor of Politics and Religion at the Faculty of Theology and Religious Studies, University of Groningen. She received her PhD in Political Science from the University of Queensland, Australia, in 2008. Her research focuses on the intersection of religion with various dimensions of politics and public life at the local, national, and global levels. She has published on religion and global justice, globalization, active citizenship, and the politics of asylum in *International Studies Quarterly, Journal of Refugee Studies, Global Society, Globalizations,* and *Politics, Religion, Ideology*. Her books include *After Secularism: Rethinking Religion in Global Politics* and *The Refugee Crisis and Religion: Secularism, Security and Hospitality in Question,* coedited with Luca Mavelli.

5

THE RIGHT CLIMATE

Political Opportunities for Religious Engagement in Climate Policy

Evan Berry

A T THE INTERSECTION OF RELIGION AND CLIMATE CHANGE, there is a strong propensity to theoretical idealism. Among scholars and journalists who write about how religion comes to matter in public conversations about climate change, there is a widespread assumption that religious beliefs and ideas are the central forces that shape how religious communities respond to the challenge of climate change. Furthermore, such considerations generally presume that religious beliefs and ideas are antecedent to their expression in social and political life—that such ideas are manifestations of history and tradition. While it may well be so that religious ideas—more specifically, theological beliefs and vernacular theologies—are critical forces in shaping the ways that various communities conceptualize and respond to climate change, the narrative frame that dominates academic and journalistic coverage of this intersection merits closer scrutiny. As conversations about the religious and spiritual dimensions of the climate crisis expand and reach new audiences, it is worth insisting on nuance and theoretical sophistication. More than a half century after Lynn White's (1967) "The Historical Roots of Our Ecologic Crisis," have we been able to move beyond oversimplified notions that underlying religious beliefs cause environmental degradation or that changing those beliefs will pave the way toward sustainability?

Mapping the Religion and Climate Discourse

In this chapter, I articulate a contrarian view, examining the conventions and assumptions in contemporary discourse about the relationship between religion and the politics of climate change. I discuss the logic within which this discourse operates and seek to identify how and why so many different voices from so many different quarters of society insist (or assume) that ideas are the driving forces in religious engagement with climate change. After mapping some of the key features of this public discourse, I focus on two prominent cases: the often noted tendency of (White) Evangelicals in the United States to deny the seriousness of and anthropogenic bases of climate change, and Pope Francis's widely celebrated encyclical Laudato Si'. These two cases suggest not that religious beliefs and ideas are unimportant but rather that a focus on beliefs and ideas can oversimplify and obscure the material, political, economic, and social dimensions of how religion comes to matter in environmental contestation.

Over the past decade, as political urgency about climate change has intensified, stakeholders from various sectors have called for religious contributions to join the global effort to tackle climate change. Although religious groups have participated in international forums on climate policy since the inception of the United Nations Framework Convention on Climate Change (UNFCCC) in 1992, their position has moved from the margins toward the center in recent years (Berry and Albro 2018). This shift is in part because religious organizations and faith-based civil society groups have become better organized and more sophisticated in their climate advocacy efforts (Berry 2014). But the public visibility of religion in contestations about climate change has also grown for other reasons. As the moral and spiritual dimensions of the climate crises become increasingly evident, and as the limitations of scientific information as a source of social transformation become painfully clear, powerful voices have championed the contributions and potential of religion.

There is a well-established trend among secular elites to call for religions to play a greater role in addressing climate change and other global environmental challenges.[1] In 2006, for example, the Climate Institute of

---

1. I here use the term *secular* cautiously. Many of the scientists, journalists, policy makers, and public intellectuals who laud the role of religion in climate action and advocacy are themselves religiously affiliated. More important, the distinction between the "religious" and "secular" aspects

Australia convened a diverse group of religious representatives, including Aboriginal leaders, Catholic priests, Protestants, Jews, Hindus, Muslims, and Buddhists, to highlight their shared commitment to the view that climate change is a moral issue. The report issued from this convening affirms that "faith communities could aid the broader dialogue on climate change by speaking the language of morality and of faith itself" (Common Belief 2006). This kind of generalized commitment to the power of religion to adequately capture the moral dimension of global warming was, throughout the first decade of the century, a regular feature of opinions and editorials and progressive organizing tactics. Amplifying an underappreciated theme of his long climate advocacy career, former US vice president Al Gore insisted that "appealing to their spiritual side" is the best way to break through the intractability of climate politics and reach American voters (Goldenberg 2009). In the run-up to COP 21, the UNFCCC gathering that produced the Paris Agreement, the tone and frequency of public calls for religious engagement on climate change intensified. Reporting on a 2014 meeting of scientists and religious leaders at the Vatican, Partha Dasgupta, an economist, and Veerabhadran Ramanathan, an oceanographer, wrote in *Science* that "finding ways to develop a sustainable relationship with nature requires not only engagement of scientists and political leaders, but also moral leadership that religious institutions are in a position to offer" (Dasgupta and Ramanathan 2014). Joining the chorus of scientists and politicians inviting deeper religious participation in the fight against climate change, a variety of public intellectuals—figures like Bill McKibben, Prasenjit Duara, and Amitav Ghosh—have made similar calls. Ghosh (2016), a novelist and literary critic whose book *The Great Derangement* has been influential in linking climate change to its cultural and colonial roots, directly praised religious engagement as "the most promising development" (159) in contemporary climate politics.

This discursive space—openings provided to religious leaders and organizations by purportedly secular public figures—has particular characteristics. Religion is not a neutral category; when it is invoked publicly as something relevant to the challenges of global environmental politics, it

---

of modern life is obscure and socially contingent. To the degree that it is an existential condition at the planetary scale, it is difficult if not impossible to disentangle discourse about climate change from teleological, eschatological, and morally normative positionalities. For additional, specific analysis regarding how climate change blurs the boundaries between religiosity and secularity, see Taylor 2009; Jenkins, Berry, and Krieder 2018; and Berry 2014.

performs specific rhetorical purposes. Unpacking the desires behind such public invocations helps explain how religion is understood as politically salient. Frustrated that overwhelming scientific evidence has failed to adequately mobilize political action against the catastrophic consequences of anthropogenic climate change, many scientists and activists have been forced to reconsider their communications strategies. To better connect with constituencies unmoved by empirical data or risk analysis, experts have recommended that climate advocates pursue a values-based approach to communicating about climate change (Corner, Markowitz, and Pidgeon 2014). Although religious voices have been part of the climate change conversation from the outset, recent efforts to foreground religion in public deliberations is part of a systematic recalibration by political elites, especially scientific spokespersons, journalists, and influential environmental advocates. Acknowledging features of human psychology, including our propensity to align with social peers and our tendency to evaluate evidence within our values-based systems of reference (i.e., cultural cognition), climate advocates have recognized the rhetorical power of ethical, rather than actuarial, discursive framings (see, e.g., Markowitz and Shariff 2012; Newman et al. 2016). In this spirit, the turn to religion represents a new attempt by environmental leaders to mobilize politically inert societies.

Those seeking to engage and advance religious contributions to climate action operationalize religion as a kind of social phenomenon that is at once institutional and ideological. The turn toward religion as a site of political coalition building is at once a departure from and a continuity with the norms of international diplomacy. Powerful religious actors have been cozy with the foreign policy establishment in the United States and other globally influential states for many decades (see, e.g., McAlister 2018 or Burnidge 2016), whereas progressive civil society organizations working in policy domains proximate to climate change, including population, sustainability, and biodiversity, have generally assumed an ambivalent posture toward religious institutions (Berry 2019). On one level, newfound enthusiasm for religious participation is part of a broader trend toward integrating religion into policy discourses (Johnston and Sampson 1995; Banchoff 2008). On another level, the turn to religion in contemporary discussions of climate change represents a novel development, an acknowledgment that the apparatus of global governance and the informational output of scientific research are insufficient mechanisms by which to affect the kind of systematic changes that will be required to meaningfully address global warming. Politics necessarily take shape within social contexts; the fact of postsecularism has found its way into the highest echelons of climate policy discourse.

This broad effort to incorporate religion into the global system of climate politics typically points toward one of two interlinking concepts of what religion is and why it matters. The first of these concepts frames religion as a moral force. In international political discourse about climate change, religion is invoked and welcomed as a mobilizer of ethical norms, a discursive domain within which the need to respond to climate change can be translated into particular concepts that are salient to culturally distinct communities around the world (Baumgart-Ochse et al. 2017). As Dasgupta and Ramanathan (2014) succinctly put it, "moral leadership will mobilize people to act" (1458). In this framing, religions are posited as the social technology through which moralities are inculcated in populations: "Religions understand the process of conviction very well. . . . The world's great religions are the winners from thousands of competing religions that managed to find the formulae for moving, exciting and persuading people" (Marshall 2015). It is beyond the scope of this chapter to critically deconstruct this notion of religion, but suffice it to say that this conceptualization aligns with secularist instrumentalism and perennialist cosmopolitanism. Latent in the view that religion serves to translate universal problems of legitimate moral concern into local vocabularies is the assumption that the work of ethical translation flows primarily in one direction. Invitations to religious leaders extended by scientists and policy makers are genuine and urgent but also are instrumentalist in that they do not envisage criticism and do not anticipate modes of religious understanding that diverge from the moral consensus about the dangers of climate disruption. The invitation to religion is an invitation to partnership in political projects already underway; it is not an invitation to prophetic radicalism or to any form of apocalypticism.

The second concept of religion operative in climate change discourse centers on the breadth and capacity of religious institutions. This conceptualization is similar to the above idea of moral translation but focuses not on messaging and ethical vocabularies but rather on the reach and power of religious networks. Religious institutions and organizations are large, powerful, and well-funded civil society groups that have the capacity to expand the global conversation about climate change to a great many individuals. As the thinking goes, "faith leaders are increasingly called upon to demonstrate leadership when speaking about the moral imperative of action . . . and guiding their congregations and communities toward the development of solutions" (ecoAmerica 2016, 7). In general, religious networks are described as important amplifiers for climate messaging, but some commentators treat religion as an essential medium for climate communications. Patrick Nunn (2017), a geographer with expertise in climate adaptation policies, makes such an argument: "The failure of external interventions for climate-change adaptation in Pacific Island communities is the wholly secular nature of their messages." In this case, the valuation of religion assumes that religious leaders are powerful gatekeepers whose cooperation is necessary to the project of broadcasting messages about climate change to a wider range of local communities.

"Religion"—understood variously as a form of moral rhetoric or as a form of transnational power—is more frequently discussed in contemporary climate discourse than it was at the outset of the twenty-first century. My aim in the preceding paragraphs has been to show that in recent years, powerful elites have warmed to the idea of including religious actors in international policy conversations about the global environment. Such welcomeness is predicated on an enthusiasm for religious engagement that imagines religion as a social space within which climate ethics can be meaningfully advanced. In such discourse, religions are promoted as repositories for values and beliefs that can be utilized in the broader project of international climate action. This observation is the basis for the two central questions of this chapter. How does the focus on "religious belief" shape and constrain contemporary conversations about the relationship of religion to climate change? And how might our perspective change if religious beliefs were not taken as the sole animating factor in religious engagements with climate change?

My concern thus far has been with the ways that political elites have shifted how religion is framed in discourse about ecological issues of planetary significance. In seeking to emphasize and catalyze the moral dimension of climate change, such elites have sought to identify and empower religious voices that resonate with axiomatic global norms. Efforts of this kind require knowledge about religious difference and sensitivity to ways in which religious difference may be environmentally salient; therefore, the effort to include religious stakeholders in climate action revisits and revivifies forms of knowledge production that capture how religion shapes environmental comportment. Following from, or at least testing, the premise of

Lynn White's critical article, environmental sociologists and sociologists of religion have long sought to determine how religious sentiment impedes or advances pro-environmental affect (e.g., Carr et al. 2012; Zaleha and Szasz 2015).[2] Research on this topic from the 1980s and 1990s was largely inconclusive, suggesting that religion was but a secondary factor in shaping public opinion about environmental issues or predicting pro-environmental behaviors.[3] Research from this era describes a complicated cultural landscape in which religion is powerfully related to environmental issues in some quarters but not in others and in which religion often serves as a proxy for complex amalgamations of social identity (see Berry 2016; Taylor, Van Wieren, and Zaleha 2016; Jenkins, Berry, and Kreider 2018). Above and beyond these lessons, this body of research indicates the conceptual slipperiness of "religion," especially when rendered as "religious belief." Studies testing the correlation between "biblical literalism" or "creation belief" and environmental concern tell us as much about broader patterns of religiopolitical polarization as they do about intrinsic relationships between theological precepts and environmental commitments.

Despite such theoretical and methodological ambiguities, the past decade has seen a proliferation of new research on the narrower question of religious belief as it shapes public opinion about climate change. Scholars and pollsters have drilled into the empirical details on this question, their curiosity driven in large part by the persistent and visible tendency of White evangelical Christians in the United States to deny the scientific veracity and anthropogenic causes of climate change (e.g., Barker and Bearce 2013; Smith and Leiserowitz 2013). Recent research on this topic suggests a more nuanced view regarding how religion comes to matter in environmental

---

2. The term *sentiment* is intentionally ambiguous: it speaks to the elusive connections of human mental and emotional states to forms of personal and collective action. This ambiguity has been treated variously by environmental sociologists and sociologists of religion, who have tested the relationships between: religious belief and pro-environmental views; religious identity and pro-environmental views; religious behavior and pro-environmental views; religious behavior and pro-environmental behavior, et cetera. The large body of literature on this topic suggests that the causal and correlative relationships among these different sociological variables are in fact more organic than mechanistic, and thus, one of the purposes of the present chapter is to decenter private measures of religiosity in favor of a social scientific explanation that prioritizes the activity of religious collectivities that operate as discrete entities within the broader socio-political ecosystem.

3. It is important to note that the data collected in this period overwhelmingly concentrates on the United States, and thus our understanding of how religious belief and environmental concern overdetermines the significance and relevance of the American example.

contestation. One line of research emphasizes the dynamic interaction between religion and political ideology, suggesting that religion is best modeled as a second-order factor that selectively amplifies partisan attitudes about climate change (e.g., Arbuckle 2017). Another line of research highlights the covalence of religion and race as they conjoin to shape environmental attitudes (e.g., Jones, Cox, and Navarro-Rivera 2014). The critical point does not concern the particular findings of such studies but instead attends to the underlying purpose served by this body of knowledge.

Empirical information about religion and climate change is grounded in a normative epistemology—understanding which cultural groups represent impediments to policy progress and which represent opportunities through which progress can be further realized. Knowing why American Evangelicals (and which ones) are climate deniers is practicable information. Knowing which religious groups have indeterminate views about climate change is practicable information. Knowing which religious communities are concerned about climate change but are less vocal or well informed about it is practicable information. Efforts to understand and explain variations among religious groups with respect to their levels of concern about climate change are parts of a broader political project of maximizing the coalition in support of climate action, internationally and in the United States in particular. This enterprise is not an illegitimate one, but the purpose of such knowledge production is not typically made explicit.

Attempts to measure religious beliefs and to quantify their significance for environmental progress are grounded in a political reality in which secular moral norms are fixed and incontrovertible; religious beliefs are to be known in such ways that they can be placed in proper relationship to the normative existential project of climate mitigation and adaptation. This conceptual position is appropriately understood as theoretical idealism— that is, an understanding of religion as beliefs and ideas (about the transcendent) that are enacted upon the (immanent) world. Within religious studies, the critique of what I am calling theoretical idealism is widely circulated and well developed as a means to understand how contemporary societies are constructed through processes of differentiation and powers of secularism (e.g., Asad 2003; Calhoun, Juergensmeyer, and VanAntwerpen 2011; Wilson 2012; Fessenden 2014). The specification of religious difference as a matter of belief, or conscience, is rooted in Euro-American traditions of thought that reflect the need to contain violence between Christian factions (see Nussbaum 2012). Much cutting-edge scholarship in the study of religion deconstructs and/or rehistoricizes how religion has been understood within bodies of knowledge about law, politics, gender, colonialism, and so forth. It is thus somewhat surprising that theoretical idealism continues to flourish in both academic and popular conversations about religion and climate change.[4] The motivation to engage religious actors on climate change stems from real and urgent concerns about the planetary future, but the forms of knowledge produced in this endeavor are manifestly part of an enduring modernist project in which the particularistic moral commitments of religious groups are distinguished from the universal moral commitments that underpin (secular) policy discourse.

Two Key Cases

How might we conceptualize the relationship between religion and climate change without lapsing into a simplistic theoretical idealism?[5] By reexamining two widely discussed cases—Pope Francis's encyclical Laudato Si' and the tendency toward climate denialism among white American Evangelicals—I use the remainder of this chapter to develop an alternative approach. By de-essentializing the role of belief in shaping these opposite forms of religious engagement in climate change, I bring to the foreground more mundane features that help contextualize how and why religion comes to matter in climate politics in particular ways at particular moments. This analysis draws on opportunity structure theory, scrutinizing not just the statements and activities of religious actors but also the political spaces within which those actions take place. Opportunity structures are the conditions within which social actors make choices, including norms, rules, means of action, and forms of incentive. For decades, political sociologists and social movement theorists have used the concept of opportunity structures to designate the "aspects of the political

---

4. Erin Wilson's chapter in this volume takes a complementary line of analysis and has been helpful in concretizing my ability to articulate this point of view.

5. In an important sense, the aim of generating nuanced, multidimensional knowledge about religion and climate change that did more than simply assert that fixed religious beliefs are realized as environmental praxis is the underlying motive for the Henry Luce Foundation funded project from which the present volume is derived. As is described in the introduction, this project sought not only to evaluate how religious sentiments shape social experiences of environmental change, but also to examine how religious actors come to play a role in climate politics and to investigate how climate driven environmental changes are acting upon religious communities, ideas, and practices.

system that affect the possibilities that challenging groups have to mo-bilize effectively" (Guigni 2009, 361). Much theoretical literature on op-portunity structures concentrates on the political conditions existing within particular states, although some recent scholarship has applied this lens to global politics and international organizations (e.g., Van Der Heijden 2006; Uldam 2013). Following these analyses, the political op-portunity structure constituted by the UNFCCC and the political elites who dominate international conversations about climate change has, with regard to religious engagement, shifted from a closed structure to an open one. This openness, however, does not explain why religious ac-tors have not mobilized en masse around the issue of climate change. Only certain groups, and those only at certain levels of institutional or-ganization, have done so. The efforts of the Holy See of Rome to support and guide global action to ameliorate climate change and the efforts of certain influential evangelical Christian groups to obstruct and obscure climate politics are both intelligible as attempts by religious stakehold-ers to maximize their political efficacy while working within the social conditions available to them. By framing the climate-advocacy efforts of these two religious groups in terms of opportunity structures, the final sections of this chapter are an attempt to historicize the ways in which religion comes to matter in environmental politics. Religious beliefs and practices, like other forms of knowing and acting, are contingent fea-tures of social systems that are shaped and reshaped over time by the practical demands of political life.

There has been an incredible amount of journalistic attention to Lau-dato Si', Pope Francis's encyclical on the global environment, and to the ways that white evangelical Christians have figured centrally in the United States' factitious climate politics. Despite the central difference—that the Vatican leadership supports robust policy responses to climate change, whereas many key evangelical leaders oppose such measures—these two cases point to a shared fundamental characteristic: there are massive incen-tives for religious actors to take a strong public stance on climate change. As climate politics have intensified over the past two decades, the concomitant public debates have become an important space within which religious ac-tors exercise their capacity for political influence. Considerable explanatory power can be leveraged by setting aside the questions about the degree to which theological commitments drive such activities and instead focus-ing on the more mundane practicalities of how, why, and when religious actors influence environmental politics. I do not argue that religious ideas are unimportant but merely assess the means by which certain kinds of religious ideas or perspectives are brought to the fore. If religious actors operate within the same systems of political economy that other kinds of social actors inhabit, then it should be possible to describe the shape and scope of religious engagements with climate change by concentrating on their interests and opportunities.

Although the basic science behind global warming has been well understood for more than a century, it remains a relatively novel policy issue, with relatively little direct attention from civil society groups dating back more than two decades. Religious actors, like other kinds of civil society stakeholders, have become gradually more and more involved with the politics of climate change during the past twenty years and have done so iteratively, along specific channels of action, within particular channels of opportunity. As with any historically specific issue, religious engagement with climate change does not operate in a space of radical openness but rather attends to established modes of power and operates within new and rapidly evolving political landscapes. My attention to these two familiar cases concentrates on the interplay between climate policy infrastructure and the political agency of religious actors.

I articulate these two cases in contrast, rather than comparison, to one another. The case focused on American evangelical Christians demonstrates how opportunity structures in the United States condition the responses of different religious constituencies. The case of the Catholic Church applies opportunity structure theory at the transnational scale, framing the Vatican's engagement with climate change in the context of its international positionality. As other chapters in this volume indicate, Catholic groups within certain national contexts may resist climate action just as many evangelical Christian networks outside the United States are positively engaged with this issue.

Evangelical Denialism in the United States

According to the Public Religion Research Institute (PRRI), nearly two-thirds (64%) of White evangelical Protestants are either "somewhat unconcerned" or "very unconcerned" about climate change and well more than one-third (39%) are skeptical that "the average global temperature is rising."[6] These tendencies are in strong contrast to other demographic subgroups: nearly three-quarters (73%) of Latino Catholics and almost two-thirds (61%) of religiously unaffiliated Americans say they are "somewhat concerned" or "very concerned" about climate change (Jones et al. 2014). A report from the Pew Research Center corroborates the PRRI findings, similarly showing that more than one-third (37%) of White evangelical Protestants in the United States agree that "there is no solid evidence the Earth is getting warmer because of human activity" (Stokes, Wike, and Carle 2015). The Pew study also shows that white evangelical Protestants are more likely than any other religious group (47%) to say that "scientists do not agree that the Earth is getting warmer due to human activity." Given that only half of Americans acknowledge that climate change is real and that human activities are its primary causes, it is important to acknowledge that climate denial is not uniquely an evangelical phenomenon. Nor are all white Evangelicals climate deniers; in fact, more than one-quarter (28%) affirm the anthropogenic causes of global warming (Stokes, Wike, and Carle 2015). Nevertheless, the tendency of White evangelical Protestants toward denialism is among the most striking and consistent correlations evident in the ample polling data on climate change public opinion in the United States. Scholarly analyses advance a variety of explanations for this strong correlation between evangelical religiosity and climate denialism, describing it as an extension of science skepticism (Ecklund et al. 2016), as the impact of "end times beliefs" (Barker and Bearce 2013), or as a reprisal of America's culture wars (Schwadel and Johnson 2017).

These framings of evangelical denialism appear frequently in media coverage, often as a way of explaining the political basis of Republican intractability on climate policy. In trying to explain why many prominent Republican leaders backed away from their previous support for climate policies between 2007 and 2012, for example, Michael Gerson (2012) editorialized that climate change has "joined abortion and gay marriage as a culture war controversy." The nomination and election of Donald Trump

---

6. I use the term *skeptical* because that is the language used in the PRRI study. Especially given the rigorous epistemological positions developed by philosophical skeptics, I find the term unsuited to describe the forms of doubt that circulate in public conversations about climate change in the United States. Denial is the appropriate term, both because of the unserious engagement with scientific literatures on climate change and because of the specific avoidance of questions concerning the anthropogenic causes of global warming.

provoked a number of deeper journalistic explorations about the role of evangelical climate denial in supporting the president's rejection of the Paris Agreement and assertion that climate change is a "Chinese hoax." It seems that the primary purpose of media coverage of this issue is to identify regressive elements and to suggest which segments of society are ripe for conversion to the climate gospel (Silk 2018). Lifting up and promoting the work of the evangelical groups that do advocate for climate action is a closely related enterprise (Jenkins 2015).

Although journalists and commentators identify evangelical climate denial as an obstacle to policy progress, much is left unsaid. Racial identity, for example, is often left out of or marginal to public conversations about this issue. Research consistently shows that climate denial is a significant feature of White evangelical communities and that this tendency does not have any real correlate among Latinos or African Americans who identify as evangelical (Pulliam Bailey 2017).[7] Evangelical Protestants of color in the United States tend to express high levels of concern about climate change and affirm its anthropogenic causes (Jones et al. 2014). This divergence between Evangelicals of color and White Evangelicals calls into question the degree to which religious beliefs or concepts are primary in shaping opinions about climate change. What accounts for these different sentiments among groups who share, at minimum, a core set of theological convictions?

Climate denialism among White Evangelicals is produced—it buttresses particular political interests, and its expression is rewarded insofar as it serves those interests. Evangelicals are a pivotal constituency, and the religious polarization of climate change is an effective way to stymy coalitional environmental politics. Despite making up just 17 percent of the electorate, White Evangelicals comprised 24 percent of all votes in the 2016 presidential election. In other words, the lens through which evangelical political engagement is refracted necessarily accords their views outsized political relevance—an optic that magnifies the relevance of their opinions about climate change. Their outsized role in electoral politics helps explain the intensity of media attention but does not necessarily clarify the nature of evangelical rhetoric about climate change. What if the kinds of theological language often invoked in opposition to climate policy were taken

---

7. For further information on ways in which race shapes the intersection of religious identity and climate opinion, the PRRI study (Jones et al. 2014) offers instructive information.

as calculated speech acts meant to resonate with White evangelical voters and not as organic expressions of the policy implications of Christian theology? What if opposition to environmental issues was shaping theological perspectives, rather than the other way around? This question has been explored in depth by Robin Veldman in *The Gospel of Climate Skepticism*, which explores the role of 'embattlement mentality' in shaping Evangelicals' attitudes toward climate change (see especially chap. 4).

Senator Inhofe, the former chair of the Senate's Environment and Public Works Committee, has repeatedly described the "arrogance" of the idea of anthropogenic climate change, assuring citizens that nature is under God's control (*Washington Post* 2015). Given the theatrical quality of the senator's political career—in January 2015, he brought a snowball into the US Capitol to demonstrate that global warming is falsifiable—it is difficult to assess the extent to which a theology of God's radical omnipotence actually shapes his policy sensibilities. This position, however, has become an increasingly common talking point among the most trenchantly conservative politicians in the United States, especially among those who seek to legitimate their views, which are far to the right of the political center, through an appeal to religious belief or identity. Scott Pruitt, Trump's first appointee to head the Environmental Protection Agency (EPA), embodies this hybridization of antiregulatory corporate politics with homespun theological justification. Throughout his career, and especially as the national media scrutinized his questionable leadership of the EPA, Pruitt routinely invoked his Christian faith as a way of explaining his radical rejection of regulatory policies. In an interview with the Christian Broadcasting Network, Pruitt described how "the biblical world view . . . is that we have a responsibility to manage and cultivate, harvest the natural resources that we've been blessed with" (Brody 2018). In contrast to other champions of the Christian stewardship model, who advocate for conservation and wise use of natural resources, Pruitt and other powerful Republicans, including Senators Inhofe and Barrasso, deploy their theological interpretation of Genesis as an argument against environmental regulations in general.[8] The perspective afforded by theoretical idealism is unsatisfactory on this point: by what measure would it

---

8. It should come as no surprise that these politicians hail from major fossil fuel producing states, like Oklahoma and Wyoming. Similar examples of the public theology can also be found in Texas and North Dakota, which are home to large populations of Christian conservatives and abundant hydrocarbon energy deposits.

be possible to evaluate whether Pruitt and others of his type construct their policy views from immutable theological principles? Much more satisfying, however, is the observation that these principles have become powerful tools for advancing their policy aims. It appears that Pruitt "polish[ed] his evangelical bona fides for years, building a bulwark of unwavering Christian support" that helped him ascend to a position of significant federal power (Palmer 2018).

Lending theological credence to denial serves to reinforce linkages between evangelical leadership and established bases of conservative political power, an alignment within which climate change is but a single aspect. Conversely, the position of evangelical climate advocates is unduly amplified by the desire of progressives to refute theologically inflected climate denial. Figures like the climate scientist and evangelical Christian Katherine Hayhoe or groups like the Evangelical Environmental Network or Young Evangelicals for Climate Action are championed by environmental nongovernmental organizations and other progressive stakeholders because they suggest the possibility of the novel kinds of coalitions that are likely needed to shift the tenor of climate politics in the United States. The American religious marketplace is highly responsive to national political dynamics—innovations within this market are where particular kinds of religious claims about climate change enter public discourse. Why assume that such innovations flow exclusively or even primarily from authentic sources beyond the reach of corporate power? It is too blunt to argue that the kinds of theological positions that have emerged within public discourse are constructed by (secular, corporate, or industrial) elites and imposed on evangelical communities. Instead, the forms of discursive formation through which such positions are constructed are never merely internal religious exchanges but always and prominently feature the political aims of allied stakeholder groups.

Laudato Si' and the Road to Paris

The Vatican's expanding engagement with climate change has perhaps been more widely discussed than any other religious engagement with climate change. Pope Francis's "Laudato Si': On Care for Our Common Home" was hardly the first papal statement on ecological issues, but it was a landmark step toward affirming the role of the Catholic Church in global environmental politics. Quite a bit has been written about Laudato Si' and its relationship to various currents of Catholic thought and tradition, and those accounts need not be replicated here (see, e.g., Dileo 2018; Peppard 2016; Deane Drummond 2016; and other contributions to this special issue of *Theological Studies*). Instead, in this section, I examine the Holy See's engagement with climate politics during the year leading up to COP 21, the high-level international meeting that produced the Paris Agreement. In this regard, what is notable about Laudato Si' is its diplomatic deployment and variegated reception across various national and multilateral contexts.

As appears to be the case with Evangelicals, theological teaching about climate change has been received quite differently among different Catholic communities. Survey data from the Annenberg Public Policy Center suggests a deep rift between conservative and liberal Catholics in the United States regarding their impressions of Laudato Si' and their sentiments about climate change as a policy issue. In the United States, conservative Catholics are more closely aligned with conservative Protestants than with their liberal Catholic peers, thus (to a limited extent) Laudato Si' has heightened divisions within the US Catholic community. As the study's authors put it: "cross-pressured by the inconsistency between the pontiff's views and those of their political allies, conservative Catholics devalued the Pope's credibility on climate change" (Li et al. 2016). In addition to internal partisan divides, the US Catholic Church is also marked by racial and ethnic division on climate change: Latino Catholics are much more likely to see climate change as a serious problem and are much more likely to see the science as settled than their White counterparts. In the US context, then, the lesson seems to be that religion is an important space where climate debates play out but that these debates are just as intense within theological traditions as they are across them.

From the Vatican's strategic perspective, such divisiveness would seem to be a small price to pay for the tremendous positive attention the encyclical garnered. The Catholic Church has had an ongoing public relations crisis dating back at least to the 2002 publication of the *Boston Globe*'s major exposé on sexual abuse in the Catholic Church and the role of high-ranking clerical officials in hiding abuses from public knowledge. The early years of Francis's papacy were marked by a variety of Vatican efforts to reframe the image of the contemporary Church. Pope Francis's humble public style and dedication to issues of poverty and economic justice captured the media's attention, not only because these were authentic features of the first Pope from the global South but also because these features reflect the strategic efforts of the Church hierarchy (Lyon et al. 2018). In direct contrast to narratives of corruption and decline that shape journalistic coverage of the ongoing sexual abuse scandals, the publication and public deployment of Laudato Si' have helped the Holy See create a new narrative about its role in the world and its enduring importance in international affairs.

Empirical data about the reception and impactfulness of Pope Francis's encyclical on climate change are scarce. Some studies suggest that the statement helped raise awareness about climate change, but other studies show that it was, like many other interventions into debates about climate change, polarizing (see, e.g., Maibach et al. 2015). The Pope's engagement with environmental issues has found uneven reception in different national contexts: in the United States, Laudato Si' has largely been received within preexisting partisan frameworks, whereas in many Catholic-majority nations—especially those with significant vulnerabilities to climate change like Peru, Brazil, and the Philippines—the Papal message has been more resonant. The enthusiasm for Laudato Si' beyond the United States also serves as a reminder of the degree to which American religious culture is a global outlier; in other anglophone nations with significant Catholic populations (e.g., Australia and Canada), Catholics are now more likely than their Protestant peers to express concern about climate change (Stokes, Wike, and Carle 2015).

Laudato Si' is an enduring document, but the encyclical cannot be understood only as a theological exercise. The text itself stands at the center of a massive, coordinated campaign to influence the international conversation about climate change—a project that involved diplomatic engagement, media promotion, and civil society outreach. The release of the encyclical was carefully timed not only to maximize its influence on the UNFCCC negotiation process in the lead up to COP 21 in Paris but also to make the papal intervention central in the international media narrative about the Paris Agreement. News that Pope Francis was developing an encyclical on climate change began to circulate among civil society groups at least as early as summer 2014, and by the time the document was formally published in June 2015, it was received into a fever pitch of media attention. In the months leading up to COP 21, Pope Francis went on a series of major international trips (including visits to Bolivia, Ecuador, and Paraguay; Cuba and the United States; and Kenya, Uganda, and the Central African Republic) where his public remarks in each nation usually focused on the need for substantive, prompt, and humane responses to the threat of climate change. Although his approval ratings have since fallen, Pope

Francis enjoyed singular popularity on the international stage—as much as 90 percent of French Catholics, 88 percent of Belgian Catholics, 84 percent of American Catholics, and 95 percent of Catholics in the Philippines held favorable opinions of the Pontiff during this period (Pew Research Center 2018; FSSPX News 2018). Taken as the signature project of Francis's papacy, Laudato Si' was a powerful tool for connecting the Pope with the global church and rebranding the Holy See in a turbulent era.

Pope Francis's choice to focus on climate change and to return to Latin America for his first international trip following the publication of the encyclical were strategic choices. These choices reflect, among other things, the Pope's experience in the Latin American Bishops Association (CELAM), which has been highly engaged with environmental stewardship and concerned with the encroachment of evangelical and pentecostal Christianity into this historically Catholic region. The theological and pastoral issues that figure centrally in Laudato Si' can be traced back to the 2007 CELAM Conference in Aparecida, Brazil (Kerber 2018). The Pope, who was in Aparecida as Cardinal Bergolio, has now carried these issues onto the global stage and done so manifestly as part of a broader project of prioritizing the concerns of Catholics in the global South. The theology of Laudato Si' matters, but the strategic deployment of those ideas has been in service to the Church's needs and interests.

There is a rich basis for Catholic engagement with climate change, and Laudato Si' makes fulsome use of theological tradition. Political practicalities, however, also figure critically in the way that the Vatican has moved to realize its climate engagement efforts. Mindful of the weakening role of the Catholic Church in European political life, the growing power of Protestantism in public life in Latin America, and the dark cloud that the sexual abuse scandal has cast over the Vatican, Pope Francis and his advisors recognized that a bold foray into climate politics would have ancillary benefits. That many world leaders and secular elites have expressed eagerness for greater religious participation in climate politics must have made that opportunity all the more inviting. As I have been arguing, theology—including mythological and scriptural resources, religious ideas about the purpose and value of human and nonhuman life, and so forth—provides the basis for religious engagement with climate change, but it is not sufficient to explain when, why, and how religious actors engage climate politics. The content of Laudato Si' is intelligible as the relatively straightforward expression of mainstream Catholic social teaching as it relates to global environmental challenges, but the timing, purpose, and implementation of the document reflects a much more pragmatic set of concerns.

Conclusion

I have outlined two generalizable observations about religion and climate politics. First, I sought to underscore cases in which political opportunity holds as much or more explanatory power than do theological ethics—that is, religion comes into play in climate politics primarily through highly contextual, historically specific moments or spaces where religious intervention is meaningful or where religious activity is specially suited to counter other forms of public reason. Second, where the concept of opportunity structures can be helpful for understanding how religious actors engage climate politics, it becomes possible to speak of what might be called "channelization"—the delimited, politically constituted avenues through which religious engagement with climate change is possible in the first place. Religious actors, according to the specific legal and political parameters within which they operate, pursue advocacy work through scripted social processes, and these processes operate according to a set of rules in which certain kinds of religious ideas may be more or less viable as political claims.

Religious actors mobilize resources (conceptual, material, ritual, etc.) from within their traditions and deploy these resources in creative ways as they confront new and emerging global issues. However, the scholars, journalists, policy makers, and activists who work at the intersection of religion and politics need to be precise in their thinking about whether the resources afforded by tradition are themselves driving forces. To situate belief or ideas at the center of an explanatory framework is to position religion outside the province of history and to fail to appreciate the particularity with which religious ideas, practices, and powers are brought to bear on worldly problems. Religious engagements with climate change are articulated within a matrix of political realities and social conditions shaped as much by secular forces as sacred ones. Beliefs and ideas guide how religious actors interact with climate politics, but where these beliefs find traction and how they are mobilized and reproduced within the horizons of possibility are very much influenced by secular elites. Religion is predicated on the prophetic transcendence of ordinary affairs. In many cases, religious engagement flows from outside the corridors of mundane politics, but these forms of religious engagement are often not the ones that draw widespread public attention.

Instead, the cases in which religious actors are linked to and powerful within established political structures are most publicly visible.

References

Arbuckle, Matthew B. 2017. "The Interaction of Religion, Political Ideology, and Concern about Climate Change in the United States." *Society and Natural Resources* 30 (2): 177–94. doi:10.1080/08941920.2016.1209267.

Asad, T. 2003. *Formations of the Secular: Christianity, Islam, Modernity*. Stanford, CA: Stanford University Press.

Banchoff, Thomas, ed. 2008. *Religious Pluralism, Globalization, and World Politics*. London: Oxford University Press.

Barker, David C., and David H. Bearce. 2013. "End-Times Theology, the Shadow of the Future, and Public Resistance to Addressing Global Climate Change." *Political Research Quarterly* 66 (2): 267–79. doi:10.1177/1065912912442243.

Baumgart-Ochse, Claudia, Katharina Glaab, Peter J. Smith, and Elizabeth Smythe. 2017. "Faith in Justice? The Role of Religion in Struggles for Global Justice." *Globalizations* 14 (7): 1069–75. doi:10.1080/14747731.2017.1325574.

Berry, Evan. 2014. "Religion and the Politics of Global Sustainability: Some Basic Findings from Rio+20." *Worldviews* 18 (3): 269–88.

———. 2016. "Social Science Perspectives on Religion and Climate Change." *Religious Studies Review* 42 (2): 77–85. doi:10.1111/rsr.12370.

———. 2019. "Climate Change and Global Religious Pluralism." In *Emergent Religious Pluralisms*, edited by John Fahy, Jan Bock, and Sami Everett, 279–301. London: Palgrave.

Berry, Evan, and Rob Albro, eds. 2018. *Churches and Cosmologies: Religion, Environment, and Social Conflict in Contemporary Latin America*. London: Routledge.

Brody, David. 2018. "Unraveling the 'Weaponization' of the EPA Is Top Priority for Scott Pruitt." *Christian Broadcasting Network*, February 26, 2018. www1.cbn.com/cbnnews /us/2018/february/unraveling-the-weaponization-of-the-epa-is-top-priority-for-scott -pruitt.

Burnidge, Cara. 2016. *A Peaceful Conquest: Woodrow Wilson, Religion, and a New World Order*. Chicago: University of Chicago Press.

Calhoun, Craig J., Mark Juergensmeyer, and Jonathan VanAntwerpen. 2011. *Rethinking Secularism*. New York: Oxford University Press.

Carr, Wylie Allen, Michael Patterson, Laurie Yung, and Daniel Spencer. 2012. "The Faithful Skeptics: Evangelical Religious Beliefs and Perceptions of Climate Change." *Journal for the Study of Religion Nature and Culture* 6 (3) (November 2012): 276–99. doi:10.1558/jsrnc.v6i3.276.

Common Belief. 2006. *Australia's Faith Communities on Climate Change*. Sydney: The Climate Institute.

Corner, A., Markowitz, E. and Pidgeon, N. 2014. "Public Engagement with Climate Change: The Role of Human Values." *WIREs Climate Change* 5 (3): 411–22.

Dasgupta, Partha, and Veerabhadran Ramanathan. 2014. "Pursuit of the Common Good." *Science* 345 (6203): 1457–58. doi:10.1126/science.1259406.

Deane Drummond, Celia. "*Laudato Si'* and the Natural Sciences: An Assessment of Possibilities and Limits." *Theological Studies* 77 (2): 392–415. https://doi.org /10.1177/0040563916635118.

Dileo, Dan, ed. 2018. *All Creation Is Connected: Voices in Reponse to Pope Francis's Encyclical.* Winona, MN: Anselm Academic.

Ecklund, Elaine Howard, Christopher P. Scheitle, Jared Peifer, and Daniel Bolger. 2016. "Examining Links between Religion, Evolution Views, and Climate Change Skepticism." *Environment and Behavior* 49 (9): 985–1006. doi:10.1177/0013916516674246.

ecoAmerica. 2016. *Let's Talk Faith and Climate: Communication Guidance for Faith Leaders.* Washington, DC: Blessed Tomorrow.

Fessenden, Tracy. 2014. "The Problem of the Postsecular." *American Literary History* 26 (1): 154–67.

FSSPX News. 2018. *The Pope's Popularity Is at a New Low in Europe.* August 26, 2018. https:// fsspx.news/en/news-events/news/pope's-popularity-new-low-europe-40189.

Gerson, Michael. 2012. "Climate and the Culture War." *Washington Post*, January 16, 2012. www.washingtonpost.com/opinions/climate-and-the-culture-war/2012/01/16 /gIQA6qH63P_story.html?utm_term=.do6ba113e4fe.

Ghosh, Amitav. 2016. *The Great Derangement: Climate Change and the Unthinkable.* Chicago: University of Chicago Press.

Goldenberg, Suzanne. 2009. "Al Gore's Inconvenient Truth Sequel Stresses Spiritual Argument on Climate." *Guardian*, November 2, 2009. www.theguardian.com /world/2009/nov/02/al-gore-our-choice-environment-climate.

Guigni, Marco. 2009. "Political Opportunities: From Tilly to Tilly." *Swiss Political Science Review* 15 (2): 361–68.

Jenkins, Jack. 2015. "More than 180 Evangelical Leaders Endorse Obama's Carbon Reduction Plan." *ThinkProgress*, July 31, 2015. https://thinkprogress.org/more-than-180 -evangelical-leaders-endorse-obamas-carbon-reduction-plan-e2ae7ecd318f/.

Jenkins, Willis, Evan Berry, and Luke Kreider. 2018. "Religion and Climate Change." *Annual Review of Environment and Resources* 43:9.1–9.24. https://doi.org/10.1146/annurev -environ-102017-025855.

Johnston, Douglas, and Cynthia Sampson, eds. 1995. *Religion, The Missing Dimension of Statecraft.* New York: Oxford University Press.

Jones, Robert P., Daniel Cox, and Juhem Navarro-Rivera. 2014. *Believers, Sympathizers, and Skeptics: Why Americans Are Conflicted about Climate Change, Environmental Policy, and Science.* Washington, DC: Public Religion Research Institute.

Kerber, Guillermo. 2018. "Church Advocacy in Latin America: Integrating Environment in the Struggle for Justice and Human Rights." In *Church, Cosmovision, and the Environment*, edited by Evan Berry and Robert Albro, 19–36.

Li, Nan, Joseph Hilgard, Dietram A. Scheufele, Kenneth M. Winneg, and Kathleen Hall Jamieson. 2016. "Cross-Pressuring Conservative Catholics? Effects of Pope Francis' Encyclical on the US Public Opinion on Climate Change." *Climatic Change* 139:367–80. doi:10.1007/s10584-016-1821-z.

Lyon, Alynna J., Christine A. Gustafson, and Paul Christopher Manuel, eds. 2018. *Pope Francis as a Global Actor: Where Politics and Theology Meet.* London: Palgrave Macmillan.

Maibach, Edward, Anthony Leiserowitz, Connie Roser-Renouf, Teresa Myers, Seth Rosenthal and Geoff Feinberg. 2015. *The Francis Effect: How Pope Francis Changed the*

Conversation about Global Warming. Fairfax, VA: George Mason University Center for Climate Change Communication.

Markowitz, E. M., and A. F. Shariff. 2012. "Climate Change and Moral Judgement." *Nature Climate Change* 2:243–47.

Marshall, George. 2015. "What the Climate Movement Must Learn from Religion." *Guardian*, April 4, 2015. www.theguardian.com/commentisfree/2015/apr/04/climate-change -campaigners-evangelism-religion-activism.

McAlister, Melani. 2018. *The Kingdom of God Has No Borders.* London: Oxford University Press.

Newman B., J. L. Guth, W. Cole, C. Doran, and E. J. Larson. 2015. "Religion and Environmental Politics in the US House of Representatives." *Environmental Politics* 25 (2): 298–314.

Nunn, Patrick. 2017. "Sidelining God: Why Secular Climate Projects in the Pacific Islands Are Failing." http://theconversation.com/sidelining-god-why-secular-climate-projects -in-the-pacific-islands-are-failing-77623.

Nussbaum, Martha. 2012. *The New Religious Intolerance.* Boston: Belknap.

Palmer, Brian. 2018. "Did Evangelical Christianity Create Scott Pruitt?" *Grist*, July 18, 2018. https://grist.org/article/did-evangelical-christianity-create-scott-pruitt/.

Peppard, Christiana Z. 2016. "Hydrology, Theology, and Laudato Si'." *Theological Studies* 77 (2): 416–35. https://doi.org/10.1177/0040563916640448.

Pew Research Center. 2018. "Confidence in Pope Francis Down Sharply in US." Pew Research Center. October 2, 2018. https://www.pewforum.org/2018/10/02/confidence-in-pope -francis-down-sharply-in-u-s/.

Pulliam Bailey, Sarah. 2017. "Why So Many White Evangelicals in Trump's Base Are Deeply Skeptical of Climate Change." *Washington Post.* June 2, 2017. www .washingtonpost.com/news/acts-of-faith/wp/2017/06/02/why-so-many-white -evangelicals-in-trumps-base-are-deeply-skeptical-of-climate-change/?utm_term =.ebaba66e2933.

Schwadel P., and E. Johnson. 2017. "The Religious and Political Origins of Evangelical Protestants' Opposition to Environmental Spending." *Journal for the Scientific Study of Religion* 56:179–98.

Silk, Mark. 2018. "A Religious Opening for Stopping Climate Change." *Religion News Service*, November 29, 2018. https://religionnews.com/2018/11/29/a-religious-opening-for -climate-change/.

Smith, Nicholas, and Anthony Leiserowitz. 2013. "American Evangelicals and Global Warming." *Global Environmental Change* 23:1009–17.

Stokes, Bruce, Richard Wike, and Jill Carle. 2015. "Global Concern about Climate Change, Broad Support for Limiting Emissions." Pew Research Center. November 5, 2015. https://www.pewresearch.org/global/2015/11/05/global-concern-about-climate-change -broad-support-for-limiting-emissions/.

Taylor, Bron. 2009. *Dark Green Religion: Nature Spirituality and the Planetary Future.* Berkeley: University of California Press.

Taylor, Bron, Gretel Van Wieren, and Bernard Zaleha. 2016. "The Greening of Religion Hypothesis (part Two): Assessing the Data from Lynn White, Jr., to Pope Francis." *Journal for the Study of Religion, Nature, and Culture* 10 (3): 306–78. doi:10.1558/jsrnc .v10i3.29011.

Uldam, Julie. 2013. "Activism and the Online Mediation Opportunity Structure: Attempts to Impact Global Climate Change Policies?" *Policy and Internet* 5 (1): 56–75.

Van Der Heijden, Hein Anton. 2006. "Globalization, Environmental Movements, and International Political Opportunity Structures." *Organization and Environment* 19 (1): 28–45. doi:10.1177/1086026605285452.

Veldman, Robin. 2019. *The Gospel of Climate Skepticism: Why Evangelical Christians Oppose Action on Climate Change*. Berkeley: University of California Press.

Washington Post. 2015. "Sen. Jim Inhofe Embarrasses the GOP and the US." March 1, 2015. www.washingtonpost.com/opinions/a-snowballs-chance/2015/03/01/46e9e00e-bec8 -11e4-bdfa-b8e8f594e6ee_story.html?utm_term=.a661d5948d58.

White, Lynn. 1967. "The Historical Roots of Our Ecological Crisis." *Science* 155 (3767): 1203–7.

Wilson, Erin. 2012. *After Secularism: Rethinking Religion in Global Politics*. London: Palgrave.

Zaleha, Bernard, and Andrew Szasz. 2015. "Why Conservative Christians Don't Believe in Climate Change." *Bulletin of the Atomic Scientists* 71 (5): 19–30. https://doi .org/10.1177/0096340215599789.

EVAN BERRY is Assistant Professor of Environmental Humanities in the School of History, Philosophy, and Religious Studies at Arizona State University. He has previously taught at American University and Lewis and Clark College. His research examines the relationship between religion and the public sphere in contemporary societies, with special attention to the way in which religious ideas and organizations are mobilized in response to climate change and other global environmental challenges. Berry is the author of *Devoted to Nature: The Religious Roots of American Environmentalism* (University of California Press, 2015), which traces the influence of Christian theology on the environmental movement in the United States. Berry recently spent a year in residency at the State Department's Office of Religion and Global Affairs as the American Academy of Religion's inaugural Religion and International Relations Fellow. He also serves as President of the International Society for the Study of Religion, Nature, and Culture.

III.

RELIGION AND THE COMPLEXITY OF PUBLIC ENVIRONMENTAL DISCOURSE

6

CONTRADICTIONS IN POLLUTION CONTROL

Religion, Courts, and the State in India

Kelly D. Alley
Tarini Mehta

CONTRADICTIONS ARE COMMONPLACE IN CLIMATE DISCOURSES IN MANY parts of the world today and often lead to stalemates in agreements to improve water conditions and water-dependent human and nonhuman life. By *contradictions*, we mean those inconsistencies in speech and action that, in dialectical logic, theory, or practice, undermine each other—that cause one entity to pose opposite propositions or to engage in practices that contest, devalue, block, or destroy the practices of others. Marx, Giddens, Harvey, and other social theorists developed the concept of contradiction as a centerpiece of the productive schemas of society, essential to the creation and resolution of crises that are political–economic and resource based. In other words, contradictions are essential to the functioning and historical change of human life.

In this chapter, we draw from the specific discussion on contradictions developed by Donald L. Donham (1999) in his applications of Marxist theory to his work in Ethiopia. We do not claim to deal with every other variant of the concept. Our objective is to test Donham's proposition that, compared with conflicts that coincide with contradictions, "conflicts that crosscut contradictions are less threatening and may even lead to social stability" (69). We look at three conflicts that have emerged among religious, legal, and governmental leaders regarding the uses of river water in India.

The values and practices we call *religious* are broadly considered aspects of Hindu tradition and are connected to legal and governmental concerns. In these conflicts, legal interventions advance assertions of human rights and, in one case, the rights of nature while incorporating religious values.

This chapter describes a variety of contradictions—some reflect deep ambivalences related to forces of production and conflicting commodity values. In this chapter, we focus on contradictions concerning the control of water resources and the land and properties surrounding water sources. We also consider divisions in the institutions of national governance—the power struggles between the executive and judicial branches and among the various levels of the judiciary. Rivers are commodified for industrial, agricultural, and urban uses, and rivers are reified in religious sentiments and beliefs as important goddesses. We look at the specific property and resource claims and interests asserted by religious organizations, legal institutions, and government departments to understand these conflicts. As Karl Marx (1939/1973) developed in the *Grundrisse*, the dual existence of the commodity as something with specific natural properties and social properties through exchange value (the specific and the general), along with its qualitative and quantitative nature, leads to opposition and then develops into contradiction. Contradiction is dialectical because the opposition is internal to the commodity and part of a developing system of production. The contradiction between use value and exchange value is mediated through the use of money; importantly, however, in the cases presented here, it is also mediated through religious rituals and festivals, even though such mediation remains predicated on the capitalist system of production.

According to Marx (1939/1973), contradictions produce more contradictions, which then create crisis. As Lawrence Wilde (1991) explains, "The failure to realise surplus value consistently manifested itself in crises, which Marx described as the collective eruption of 'all the contradictions of bourgeois production.' But crises were not the end of the story. Marx saw them not simply as a manifestation of contradictions but also as a reconciliation of them 'by the violent fusion of disconnected factors'" (286).

Following this perspective, we consider river pollution and climate change as crises of capitalism. We look to the ways contradictions emerge from the fault lines of society, from the fundamental inequalities in resource, property, and institutional controls. We see that religious, legal, and governmental leaders are producing values that support or conflict with these inequalities. As Donham (1999) notes, productive inequalities occupy a central place in Marx's thought because they locate the basic divisions within any society, the lines of potential opposition—of contradiction. Donham explains, "Marx saw these as the potential fault lines along which tensions tend to build up, are routinely dissipated by small readjustments, and are sometimes violently resolved by radical realignments. These fault lines are structural; they do not necessarily lead to actual struggle and conflict (indeed, the function of the superstructure is precisely to prevent such occurrences). Nevertheless, contradictions always exist as potentialities; they lie just below the surface" (64).

In this chapter, we examine the legal arguments that emerge from conflicts regarding water resources. They occur among religious, legal, and governmental leaders who work against each other and sometimes with each other to procure access to rivers or to prevent access to rivers for their own needs or to promote their values. We are intrigued by Donham's (1999) proposition that some kinds of conflicts end up leading to stability or perpetuating the status quo. Donham writes:

> Conflicts that coincide with contradictions are, obviously, the most dangerous for social reproduction. For those cases, if societies are to persist in the same mode of production, methods of resolution must exist that will uphold the power of dominant groups. Exactly how this is accomplished is often a complex matter and varies from outright denial of rights to certain categories of persons (slaves in slave-based societies), to the formal exclusion of categories of persons from participation in legal procedures (women, even those directly involved, are not allowed to participate in divorce cases in some technologically simple societies), and finally to the de facto exclusion of still other groups from recourse to the law (workers in capitalist societies are generally too poor to fight certain legal matters). Conflicts that crosscut contradictions are a good deal less threatening and may even contribute to social stability. For example, lords (allied with peasants) fighting other lords (allied with other peasants) can do much to prevent the contradiction between all lords and peasants from becoming too visible, from erupting into overt conflict. Accordingly, such kinds of conflicts may resist methods for settling them longer than those discussed above. (69)

The following descriptions show that in the first case, legal leaders and petitioners oppose religious leaders and governmental leaders over water uses; in the second case, legal leaders align with religious values against governmental leaders; and in the third case, governmental leaders align against legal leaders and religious leaders. These crosscutting oppositions do not threaten or alter the fundamental fault lines of control over water or administrative and institutional powers. Because these conflicts among elite leaders do not lead to any radical changes in resource use or institutional or rights-based powers, they do not create transformations that lead to pollution prevention and water conservation.

After analyzing the dynamics of these conflicts, the concluding section of this chapter addresses whether these conflicts have produced any minor changes in religious values and practices or in legal values and practices (in terms of extending the definition of human rights to include nature's rights) that could eventually improve river quality and slow down climate change. We conclude that contradictions in terms of the oppositional positions over water uses and values do not produce structural changes in resource controls; therefore, they do not reduce pollution or overexploitation of water. This conclusion affirms Donham's (1999) proposition and explains why stalemates occur when conflicts crosscut contradictions. Without fundamental changes in resource, property, and institutional relations and controls, legal conflicts cannot prevent the ongoing contamination of surface waters or even the destructive effects of climate change.

Methodology

In the following sections, we examine the conflicts that emerge as these leaders interact with each other in the pursuit of river uses, as they use the water and river bank environs for specific purposes or attempt to protect water and floodplains from pollution and ecological degradation. Observations and generalizations made in the chapter are drawn from Alley's thirty years of fieldwork in the Ganga and Yamuna river basins of India. This fieldwork has involved focus groups; interviews that have been tape-recorded and transcribed; surveys; and participant observation among religious, political, and governmental leaders in cities and towns along these rivers and legal practitioners in the Supreme Court, high courts, and National Green Tribunal (NGT). Interview data available in the public domain, media reports, governmental and nongovernmental documents, and legal petitions, orders, and rulings are used as sources of data in this chapter.

Background

The Ganga and Yamuna rivers are featured in this chapter. They are two of the longest rivers in India and provide water to irrigate the vast area of the Doab, the farming land lying within the Ganga basin. They also provide water for growing urban centers and for the many small and large industrial centers in the north. For Hindus, both the rivers Ganga and Yamuna are goddesses. Ganga is a mother, goddess, purifier, and sustainer of all life, and Goddess Yamuna is her sister, responsible for love and sustenance in this world (Alley 2002; Alley and Drew 2012; Eck 1982*a*, 1982*b*; Haberman 2006; Kumar and James 2013).

Given the values of the rivers in terms of sacred power, purification, and love, the first great contradiction is that these rivers are severely burdened by untreated sewage. To address this severe problem, the rivers Ganga and Yamuna and their tributaries are centrally featured in river cleanup programs and advocacy (Alley 2002, 2014, 2016; Alley, Barr, and Mehta 2018; Eck 1982*a*, 1982*b*; Haberman 2006; Markandya and Murty 2000; Rauta 2015; Tare and Roy 2015). In the latest iteration of river cleanup—called Ganga rejuvenation—the Clean Ganga mission has become a rallying cry for the nation, promoted by India's Prime Minister Narendra Modi. The mission title, Namami Gange, means "obeisance to the Ganga river." There are many cases focusing on pollution of the rivers Ganga and Yamuna, and occasionally the justices on the NGT, like those on the Supreme Court, will combine or "club" cases with similar pleas and purviews to hear them as a collective. This chapter describes three cases focusing on the pollution of the Ganga and Yamuna rivers that involve conflicts among religious, governmental, and legal leaders in India, as they address the state of the country's sacred rivers.

The cases unfold within three different tribunals or courts in India. The first takes place in the NGT in Delhi, the second in the high court of Uttarakhand, and the third as an appeal of the second case in the Supreme Court of India. The NGT was established in 2010 after years of discussion about the best ways to deal with environmental issues and to dispense environmental justice in India. It is composed of judges and expert members with knowledge of science and the environment, and the expectation is that they decide matters based on environmental merits and citizen concerns. The majority of the cases taken up by the tribunal are related to pollution (31%) and environmental clearances (35%). But even though the tribunal functions to fill knowledge, training, and regulatory gaps that exist across environmental governance, some of their judgments are scrutinized and considered unsound.

The Supreme Court, once the primary court for environmental litigation through a legally unique "public interest cell," has ceded most of its environmental decision-making to the NGT, whose legal deliberations include scientific assessment and monitoring (Amirante 2012; Chezhiyan

2008; Kumar 2016; Shrotria 2015). In 2016, the Supreme Court passed one of the most important environmental cases to the NGT permanently, ceding to the tribunal its close monitoring of pollution. This marks a major shift in the strategy of *continuing mandamus*, a remedy used by citizens rather effectively in the early phases of public interest litigation in the Supreme Court.[1] In the NGT, the caseload was led by the persistent complaints of citizens, and the chairperson assumed the practice of calling stakeholders forward to deliberate the problems. In prior Supreme Court cases, the remedy of continuing mandamus was used to extend court-directed monitoring of compliance with orders and laws; it was considered necessary by judicial authorities who saw the government wasting national assets. When the Supreme Court agreed to cede responsibility to the tribunal, the media accused it of "passing the buck." The court responded, "For us, to monitor week after week, months after months, it is difficult. . . . [You] can go there [NGT] and if you have a problem you can come back to us" (see Nair 2017).[2] At the end of April 2017, the Supreme Court passed another key case on the Yamuna to the NGT (*Times of India* 2017). As one advocate explained to me, "When cases are sent by the Supreme Court to the NGT, the NGT is more motivated to hear them."

This chapter shows that some religious and governmental leaders may at times assume that they are exempt from the legal orders that apply to them as they conduct their resource uses according to the interests of their organizations and departments. They appear to engage in behaviors that break environmental policies, laws, and institutional rules relating to conservation and pollution prevention without penalties or punishments. These exemptions become noncompliance practices that we see as contradictions indicating the fault lines of water control and property holding. These noncompliant behaviors and contradictions encumber efforts to rectify water pollution problems.

---

1. Mandamus is a judicial remedy in the form of an order from a court to any government, subordinate court, corporation, or public authority to do some specific act that the body is obliged under law to do and that is in the nature of public duty and, in certain cases, statutory duty (see Alley 2009).

2. In the same article, the Supreme Court stated through the media: "The issue relating to the river development and Ganga rejuvenation, including discharge of municipal waste and industrial waste, is already being heard by the NGT. So, we transfer the instant petition to the tribunal and the entire proceedings pertaining to the cleaning of the river shall be placed before the NGT" (Nair 2017).

Religious Foundation, the State, and the NGT

We begin with a recent case about the Yamuna river that involves a popular religious leader. This section will take a deep look into the relations among religious, legal, and governmental leaders. This case involves a large temporary construction by the Art of Living Foundation (AOL) on the floodplain of the Yamuna river in Delhi. The leader, Sri Sri Ravi Shankar, is an established spiritual guru with more than 370 million followers around the world (AOL 2021). Sri Sri Ravi Shankar established the AOL in 1981 to perform humanitarian work with a focus on "stress-free" living. The foundation runs hundreds of centers where people take classes on yoga, personality development, the achievement of happiness, and healthy family relationships. The foundation also has a successful line of Ayurvedic products. The leader Shankar participates in current affairs and politics by attending important events and engaging in discussions with political leaders. He is able to influence large transnational audiences through projects, visits, and lectures and attracts intense media attention (Ani 2016). His speeches usually favor the current National Democratic Alliance government and Prime Minister Narendra Modi.

In early 2016, AOL decided to host a massive devotional event in Delhi called the "World Culture Festival." The event took place on the floodplain of the Yamuna river at an unusually vacant part of the overcrowded and tightly urbanized capital. The AOL planned and constructed temporary ramps, roads, and bridges to facilitate the event. It used approximately 400 hectares, a land area greater than the space originally granted by the Delhi Development Authority (DDA). To expand the project space, the AOL encroached on farmers' lands (Wilkes 2016).

The foundation also carried out its own waste disposal practices. Before the event, Shankar and his staff had to address the stink of the Barapullah nala (drain) that ran adjacent to the festival grounds. The gaseous smell would be off-putting and even dangerous to visitors' health. Thinking of ways to tackle a wastewater flow that exceeded 100 million liters each day, foundation members approached a local nongovernmental organization (NGO) with experience in the bioremediation of wastewater in drains and lakes.[3] The NGO declined to work on the project without proper government permissions, so the AOL representative approached the DDA for

---

3. Bioremediation is a form of treatment that uses natural processes including enzymes or bacterial agents to jumpstart anaerobic or aerobic digestion.

help.[4] Together, they hired a Delhi-based company to carry out bioremediation of the drain. They dripped a bacterial agent into the sewage flow to jumpstart the process of digestion, reduce the odor, and bring down the biological oxygen-demanding content. The AOL then instructed devotees to help in the process by adding the bacterial agent to sugar cane juice and pouring the concoction into the drain.

When Manoj Misra, a local environmental activist, heard about their plans, he went to the NGT with a petition to shut down the event. As convener of the Delhi-based NGO Yamuna Jiye Abhiyan (Long Live Yamuna movement), Misra claimed that AOL had not procured any permission to drip a bacterial agent into the drain and was not disclosing the nature of the bacterial agent. Immediately, the NGT ordered the organization to cease dripping the agent into the drain. The petitioner and the tribunal bench then began an inquiry into the nature of the AOL's activities and permissions. The NGT eventually issued a stay on the bioremediation activity, and the DDA instructed the company to discontinue the applications.[5]

On March 10, 2016, the NGT declared that the AOL's World Cultural Festival had caused "damage . . . to the environment, ecology, biodiversity and aquatic life of the river" and "should be held liable for its restoration in all respects."[6] Four hundred hectares of land had been used for roads, ramps, pontoon bridges, and other semipermanent structures without the permission of concerned authorities. The permissions granted by the Government of Delhi were also not sufficient for the activities that the AOL pursued. Its construction activities violated the principles laid down in other legal cases on Yamuna and the floodplains.

The AOL did not seek environmental clearance for any aspect of the event or for drain cleaning or construction. In its defense, the Ministry of Environment, Forests and Climate Change later stated that the AOL did not need environmental clearance.[7] In another previous case, the NGT had

---

4. The DDA was set up under section 6 of the Delhi Development Act of 1957.

5. *Manoj Misra v. Delhi Development Authority & Ors., and Art of Living International Center and Ors.* (Original Application No. 65 of 2016).

6. *Manoj Misra v. Delhi Development Authority.*

7. *Manoj Misra v. Delhi Development Authority.* Reprimanding the ministry, the NGT pointed out that its stand was contrary to the law: "We also do not accept the contention of the MoEF & CC [Ministry of Environment, Forests, and Climate Change] that it was not required for the Foundation to seek Environmental Clearance for the project relating to all matters of construction etc. as afore-referred. The stand of MoEF & CC is contrary to the Notification, particularly with respect to development of an area of more than 50 ha. as contained in the EIA Notification, 2006."

articulated principles of floodplain protection. Related to these principles was the prohibition on construction activity in the demarcated flood plain.[8] The committee constituted by the NGT in the AOL case made note of the fact that the event and the permission given to it by the DDA violated previous orders of the NGT. Making special note of the judgment of the court in *Manoj Misra v. Union of India* (2012), the committee stated that "such incidents should not happen ever again and should not be tolerated in the least."[9] It found that the AOL violated several environmental norms and created large-scale damage before the completion of construction. The damage would require substantial and well-planned eco-restoration measures.

Despite AOL's disregard for the law and for previous judgments of the tribunal, the NGT did not prevent the event from continuing. The tribunal had started examining the damage just a month before the scheduled event and, at that time, issued notices to the DDA, the Delhi government, and the AOL. It did not issue a stay during the adjudication of the matter. In its final judgment on March 10, 2016, the NGT stated that the excessive delay created a situation in which it was too late to prohibit the event: "For the reason of delay and laches on the part of the applicant in approaching the Tribunal and for the reason of fait accompli capable of restoration and restitution, we are unable to grant the prayer of prohibitory order and a mandatory direction for removal of construction and restoration of the area in question to the applicant at this stage."[10] The committee had calculated ecological costs to be in the range of Rs 100–120 crores ($14–16 million) and recommended that, in accordance with the "polluter pays" principle, the amount should be recovered from the AOL.[11] Before the committee could issue its report, the tribunal demanded that the AOL pay initial "environmental

---

8. *Manoj Misra v. Delhi Development Authority.* The court held, "Subject to any law coming into force, we have already stated that flood of once in 25 years would be considered for defining and demarcating the flood plain. No development/construction activity, except that is stated herein, would be permitted in the Flood Plain of River Yamuna. No authority or person before us has even taken up the plea that why development/construction activity cannot be carried on in other parts of NCR, Delhi. As of now, sufficient land is available, may be it is expensive, but that cannot be a ground for destroying the ecology, environment and biodiversity of River Yamuna of Delhi. The result of indiscriminate, unregulated and uncontrolled development activity are widely visible and felt by each and every one in Delhi. It would not only be unwise, but may prove fatal, if such approach is continued any further."

9. Report of the Committee Named by the Hon'ble NGT for Visiting the Site of the World Culture Festival in the Matter OA No. 76 of 2016 (M.A. No. 144 of 2016), pars. 1 and 4.

10. *Manoj Misra v. Delhi Development Authority.*

11. Report (M.A. No. 144 of 2016), pars. ii and iii.

compensation" of Rs 5 crores (approximately $685,000). After a great deal of protest and delay, the foundation paid the initial compensation amount (Majumdar 2016). Then, in the report, the seven-member committee stated that Rs 42.02 crores ($5.7 million) was required for ecological rehabilitation of the area. They recommended that the costs be divided between the AOL and other agencies.

In the following hearings, the tribunal chairperson and Ravi Shankar exchanged threats. The tribunal blamed him for the damage to the flood plains. He responded that the fine should be imposed on the NGT, not on him, for permitting the event. He posted on his Facebook page, "If, at all, any fine has to be levied, it should be levied on the central and state governments and the NGT itself, for giving the permission. If the Yamuna was so fragile and pure, they should have stopped the World Culture Festival." The guru went on, "World over, cultural programmes are held on riverbanks. The whole idea was to bring awareness to save the river. The Art of Living that has rejuvenated 27 rivers, planted 71 million trees, revived several ponds is being projected as destroying a dead river. What a joke" (Thakur 2017).

This comment led the chairperson of the NGT, Justice Swatanter Kumar, to tell Ravi Shankar, "You have no sense of responsibility. Who gave you the liberty to speak whatever you want to? It is shocking." He accompanied this with a contempt notice to the foundation, but the foundation denied receiving it. When asked about the case, the petitioner said, "At least the NGT have done something, they have abused these people. I don't think anyone else would have done even that" (Thakur 2017).

In its earlier decision in *Manoj Misra v. Union of India*, the NGT required that concerned authorities "educate all sections of society to co-operate and not to do any acts or deeds which are prohibited under this judgment and would have adverse consequences." They were to "issue circulars, display signages and take recourse of print and electronic media for educating people at large for effective completion of this project."[12] The point of this language was to stress the need to spread awareness regarding the conservation of the Yamuna floodplain and to generate respect for environmental laws. The principal committee set up by the NGT wanted a strong message sent by this case to "prevent any attempt for further such violations

---

12. *Manoj Misra v. Union of India & Ors.* (Original Application No. 6 of 2012), as per judgment dated 13-01-2015, par. xvii.

in the future."[13] However, the weak decision in the AOL case sent a completely different message. It indicated that the preservation of ecology and the rule of law could be set aside if the people involved were influential. The NGT case against the AOL makes clear that a religious leader can avoid compliance with prior principles on floodplain protection and can delay or evade tribunal orders, creating contradictions in the power of the law. This case illustrates the tensions that developed through legal conflicts as the NGT, following the tradition of continuing mandamus, assumed the role of a monitoring agency. The NGT has to assume the regulatory role that should be performed by government departments, such as the Ministry of Environment and Forests and the pollution control boards. The AOL case demonstrates that the tribunal is willing to set aside environmental legal principles, and previous decisions, to issue mild punishments and fines. If it had followed its judgment of January 31, 2015, the NGT should have been able to prevent the World Cultural Festival from constructing its stages on the banks of the Yamuna. However, the tribunal was unable to enforce its own decision or impose punishment for other irregularities in granting permissions to land and water uses.

Returning to the focus on drains, Sri Sri Ravi Shankar did not have any long-term interest in fixing the sewage problem in the Barapullah nala or finding solutions for the ongoing pollution of the river Yamuna, which, by all ecological measures, is severe. The AOL's actions show that some religious leaders have no abiding interest in finding solutions for the terribly septic state of the drains that feed into the country's rivers. In this case, the religious group's interest was in the appeal of its cultural event and the symbolic and political capital it could generate.

This case exposes the religious organization's interest in commodifying the river by using its sacredness as symbolic capital for its event. This legal conflict is also one of several involving the use of government land, floodplains, and other vacant land for religious organizational activities. In another instance, a case was filed in the High Court of Chennai against the Isha Foundation, a religious organization led by spiritual guru Jaggi Vasudev. In that case, the petitioner claimed that illegal constructions had been set up in an ecosensitive wetland area of Coimbatore, Tamil Nadu. The petitioners were seeking demolition of a 1.3 million sq ft construction built above a canal in the wetland on property held by the government under the Hill Area Conservation Authority. Despite the objections of local citizens,

---

13. Report (M.A. No. 144 of 2016), Specific Recommendation v.

in addition to the pollution control board and the district forest officer, the district collector of Coimbatore failed to halt the illegal construction. The high court also delayed hearing the litigation (Lakshmana 2017).

In another NGT case in 2017, a petitioner sought an interim order to prevent the Isha Foundation from conducting the Mahashivratri festival due to the adverse impact it would have on the elephant corridor. The majority of the land owned by the Isha Foundation is located in the foothills of the Velliangiri mountains and is part of the elephant corridor. The petitioner argued that the increased noise levels and high numbers of attendees (around 2 lakh [200,000]) would adversely affect the wildlife of the area. The district forest officer also objected to the festival. Although the tribunal admitted the petition, it refused to pass an interim order and instead ordered a further hearing on the matter. Consequently, the festival went on without an obstacle. Like the AOL function, the event was attended by several dignitaries and political figures, including the prime minister, who inaugurated a 112-ft statue of Adi Yogi. Environmental groups asked the prime minister not to attend, as that would condone the environmental violations (see Iyer 2017; Purkayastha 2017; Ramasubramanian 2017). A former Madras high court judge also condemned the prime minister's visit, stating that "the Isha Foundation has over 13 lakh square ft of illegal constructions . . . the visit by the Prime Minister may interfere with the administration of justice" (Purkayastha 2017).

Legal Leaders, Government Leaders, and Religious Values: Ganga and Yamuna as Persons

Another set of contradictions occurred in the case that produced the high-profile ruling on Ganga and Yamuna as juristic persons. This case was heard in the high court of the state of Uttarakhand. It does not involve the NGT, even though the subject matter was within the tribunal's purview. The case was originally filed as a grievance about encroachment along one tributary of the Ganga canal that supplies fresh water to farms, industries, and cities in the Gangetic plain. Over time, the case migrated from its focus on territorial disputes between occupiers and evictors to an interest in the identities of sacred rivers. The change in judges hearing this case also accounted for the shift in the direction of the case.[14]

---

14. Bhuwania (2017) calls this a trend of "omnibus PILs," in which a case begins in court as one thing and ends up as quite another (see Shivshankar 2017).

On March 20, 2017, the High Court of Uttarakhand ruled, "The Rivers Ganga and Yamuna, all their tributaries, streams, every natural water flowing with flow continuously or intermittently of these rivers, are declared as juristic/legal persons/living entities having the status of a legal person with all corresponding rights, duties and liabilities of a living person in order to preserve and conserve river Ganga and Yamuna."[15] The immediate motivation for the landmark ruling, as Lokgariwar and others have noted (Gopalan 2017; Lokgariwar 2017), comes from the case of the Whanganui river in New Zealand.[16] In that case, the river was declared a living entity with full legal rights by the country's parliament after a long push by the Maori, an Indigenous group. In India, a discussion about the need to grant personhood to these rivers was introduced a year earlier by the Community Environmental Legal Defense Fund and then shaped into a declaration by a group of twenty-five religious leaders of different faiths. In September 2016, this group called on the government to declare Ganga a person as a way to enforce stronger steps to protect her (*Financial Express* 2016). Led by the head of a popular ashram in Rishikesh, this collective submitted its letter of request to the union science and technology minister at an event organized by Global Interfaith WASH Alliance and UNICEF-India. It said a new law should be enacted to "grant legal rights to the Ganga." The leader of this group noted to the media:

> "We feel that by enabling the Ganga to secure its own legal rights to survive and thrive, just like a company or a person, the same can later be done for all other rivers such as the Indus," the religious leader said. On the need for a new law despite one already being in place, he said the legislation, Water (Prevention and Control of Pollution) Act, 1974, has been "rendered dry" over the past four decades. In 1974, a Water Act was passed that called for up to seven years jail for repeat polluters. Yet 42 years later, our nation is in trouble. "80 percent of our drinking water is polluted, mostly with sewage. The Water Act is a dry piece of legislation," he said. (*Financial Express* 2016)

While this leader attributed the problem to the standing law and called it "dry," others argue that compliance is the more critical issue. In the ruling, the high court judges appointed three guardians—the director of Namami Gange, the chief secretary of the state of Uttarakhand, and the

---

15. Writ Petition (PIL) No. 126 of 2014, *Salim v. State of Uttarakhand*, para. 19.

16. The direct inspiration from the transnational discourse on rights of nature and this ruling is not explicit in the judgment but was communicated during interviews in October 2018. The advocate cited the New Zealand case in the hearing, and one judge followed up by exploring the literature and cases on rights of nature from around the world (see Alley 2019).

advocate general of the state—as "persons in loco parentis as the human face to protect, conserve and preserve the Rivers Ganga and Yamuna and their tributaries." By this order, the officers were bound "to uphold the status of [the rivers] and also to promote the health and well-being of these rivers" (Gopalan 2017). The court confirmed that any harm done to the river would be a cognizable offence and the state would initiate criminal proceedings without waiting for a petitioner. Ten days later, as people were still trying to understand the implications of this order, the court, in another case, designated the glaciers, lakes, and wetlands of these basins as "legal persons."[17]

Although the judgment is ostensibly for environmental conservation and protection, its roots in religious sentiments and philosophy cannot be set aside. Justice Sharma began his presentation on these aspects by stating, "Rivers Ganges and Yamuna are worshipped by Hindus. These rivers are very sacred and revered." He continued, "The Ganga is also called 'Ganga Maa.' It finds mentioned [*sic*] in ancient Hindu scriptures including 'Rigveda.'"[18] He used this statement as he introduced other cases in which Hindu idols or deities had been seen as juristic entities capable of holding property.

Hindu views pertaining to deities are foundational to this judgment and others, particularly in terms of the personification of different elements of nature as gods and goddesses. Of the six Astika schools of Hindu philosophy, Vedanta is one of the most popular and influential. It has three schools of thought—dvaita, vishishta advaita, and advaita—and all invoke different forms of personhood for divinity. Dvaita is the philosophy of dualism, Vishishta Advaita is qualified monism, and Advaita is monism. In terms of the spiritual journey, these three can be seen as epitomizing the progression of the consciousness toward the realization of ultimate reality, epitomized by Advaita philosophy. Followers of Dvaita and Vishishta Advaita believe in a personal god and explain devotion as the path to liberation, whereas Advaita metaphysics holds that the world has no existence separate from Brahman, the ultimate reality. The self that is experiencing,

---

17. Writ Petition (PIL) No. 140 of 2015, *Lalit Miglani v. State of Uttarakhand*, which declared glaciers, including Gangotri and Yamunotri, rivers, streams, rivulets, lakes, air, meadows, dales, jungles, forests wetlands, grasslands, springs and waterfalls as legal persons (Lokgariwar 2017; Shivshankar 2017; Studley 2018).

18. Order dated March 20, 2017, in Writ Petition (PIL) No.126 of 2014, *Mohd. Salim v. State of Uttarakhand and others*, para, 11.

or the *jiva*, and the universal self, or *atman*, are the same; they are both Brahman, despite the fact that the individual self seems distinct. The Brahman of Advaita Vedanta advocated by Sankara is impersonal. It transcends all attributes and thoughts. As described by Sankara, "Brahman [ultimate reality] is without parts or attributes . . . one without a second" ([traditional attribution], second half of the 8th century; quoted in *Stanford Encyclopedia of Philosophy* 2012). This Nirguna Brahman is the ultimate reality as it truly is. However, it becomes a personal god because of its relationship with the principle of Maya. The Advaita Vedanta school distinguishes between Nirguna Brahman (attributeless) and Saguna Brahman (with attributes); Saguna Brahman is illusory.

The abstract philosophy of advaita is difficult to grasp and comprehend; therefore, the value of the concept of a personal god is acknowledged. The philosophies and spiritual practices of Dvaita and Vishishta Advaita are, for their part, grounded in the belief in a personal god. In the philosophy of Vishishta Advaita, Brahman is a personal god, a supreme being (*Param Atman*) and creator of a world permeated by his essence (Van Buitenen 2021). Madhava, founder of the Advaita School, similarly identifies Brahman with a personal god, as Ramanuja (approximately 1050–1137) had done when expounding the Vishishta Advaita philosophy. For theistic Vedantins, devotion toward a personal god becomes the cornerstone of the path to salvation and the foundation of the bhakti tradition.

Depending on the tradition one adheres to, deities are either concepts beneficial for spiritual progress or real beings toward whom one surrenders with devotion. Either way, they are a key element of Hindu philosophy and religious practice, and deifications to natural elements and qualities are recognized as valuable for human life and society. Personhood of deities is a key aspect of the living spiritual traditions of India (see discussions by Nelson [1998] and Jain [2011] and essays on Hinduism and ecology in Nelson [1998] and Chapple and Tucker [2000]).

In relation to this personification, broadly speaking, two different approaches can be identified: one considers deities to be personifications of abstract energies and qualities, and the other sees them as real beings embodying divine energies and qualities. Both views result in worship, with the difference that, in the former case, the spiritual aspirant's worship is directed toward the energy symbolized by the deity. In the latter case, the deity epitomizes the energy or quality and is the object of devotion. In both cases, the representation of the deity is of importance, as emotional connections develop with that form and the narratives around it. During the Hindu festival of Diwali, for example, millions of Hindus pray to the goddess Lakshmi and light lamps to urge her to enter and bless their homes—firm in the belief that, through proper worship, the goddess would be propitiated and pleased. Ganga is another such deity. Worshipped as a goddess, places of worship have grown along her banks, including the great holy towns of Varanasi and Haridwar. For the devout Hindu, she is mother, protector, redeemer, and one who purifies consciousness.

In the worship of Ganga, one finds the two approaches commingling, for Ganga is at once "the Supreme Shakti of the Eternal Shiva" (Eck 1982b, 219) and also a real being. Several mythological stories about Ganga as the personified goddess exist in scriptures. One story states that she was one of the daughters of King Himavat and Queen Menavati. Another legend tells that Ganga was devoted to Lord Krishna in his divine abode. This made Radha jealous, and she cursed Ganga to drop down to earth and flow as a river. Devotees bathe in her waters to be cleansed of their sins; the ashes of the dead are immersed in her waters and lead departed souls to higher birth; and her name is chanted with the belief that it will bestow freedom from poverty and protection, even lead to liberation. For the Hindu mind, Ganga is supreme among rivers, an archetype of sacred water. Other rivers are said to be like the Ganga—others are said even to be the Ganga. Such is the strength of the belief in the personhood of the Ganga and in her divinity and powers. Yamuna, Godavari, Saraswati, Sindhu and Kaveri are the other sacred rivers of India. Two of the most sacred Hindu pilgrimage places, Gangotri and Yamunotri, lie at the headwaters of these sacred rivers. Goddess Yamuna, like Ganga, has been personified. In Hindu mythology, Yamuna is the daughter of Surya (the sun god) and Saranyu. The lord of death, Yama, is her brother (Haberman 2006; Kumar and James 2013).

Uma Bharti, when she was minister of water resources, put her own sentiments this way at an opening statement for a climate conference in 2016:

> I never looked at Ganga from the religious point of view. Because the Hindu point of view is such that we look at everything with a religious point of view. So the religious point of view is always hidden there. It is always there. It cannot be without it. A Hindu vision will never be without a spiritual vision. Trees, stones, rivers, animals, insects, stars, sky, water, air. God is existing in everything, everywhere. Ganga is the very cool economic flow of this country. Because Bihar and UP [Uttar Pradesh], the biggest populated states, are

completely dependent totally on Ganga. And Ganga is a story of how we destroy rivers. So Ganga becomes a model of how we save rivers. In that ecological flow is necessary, cleaning of Ganga through various methods is very necessary. And saving the rivers of this country because I always say that rivers and women have to fight for their own existence. Nobody helps them. They create their own existence, they save their own existence. It is very difficult for them. They struggle a lot. So the flow of women's growth and the flow of the river also.

Following the order, there was some skepticism that designating personhood to rivers based on religious beliefs would lead to better management and conservation (Alley 2019). Institutionally, this judgment would not bring any major reforms and would only add to the overbureaucratization of water in India. Some argued that there would not be any additional benefit by making the director of Namami Gange, the chief secretary of the state of Uttarakhand, and the advocate general of the state "persons in loco parentis as the human face to protect, conserve and preserve the Rivers Ganga and Yamuna and their tributaries." The Namami Gange and other agencies responsible for conserving these rivers were already failing miserably; adding an additional title did not seem to be the way to affect much change.

To understand the motivations and import of the court's decisions in March and, before that, in December 2016, we must explain center–state politics and the recalcitrance of the state during the two years preceding this ruling. The state political party, the national party, and the union and state government departments in charge of water, sanitation, and engineering constitute the major power groups in control of river water decisions, and fault lines were dividing these government groups based on state versus center affiliation. Government groups are also separated from religious organizations and citizens in terms of these powers, as they hold most of the important decision-making levers. Consequently, religious organizations, industries, and citizen monitors try to wrest power away from them or find a way to influence them.

The government tries to bring the populace along with its water decisions and allocations by appealing to religious values and claiming to respect devotional practices of bathing and using sacred water (Ganga jal) in rituals and bathing practices. The prime minister has twice contested his seat from the place of great devotional fervor, the sacred pilgrimage city of Varanasi, and when doing so claimed Ganga cleanup as his chief mission.

His first contest in the state elections—the one that brought him to power as prime minister for his first term—provoked a contest with the ruling chief minister of the state, who represented another political party, the Samajwadi Party. Before this contest, the Samajwadi Party, as a minority party in the central government, had been dragging its feet on the central government river cleanup activities under the Namami Gange initiatives. The party also appeared to be misusing central government funds. The dynamics changed after 2014 when Narendra Modi won the Varanasi seat, and the Samajwadi Party lost its majority in Uttar Pradesh. When the Bhartiya Janata Party (BJP) represented by the Prime Minister took control in Uttar Pradesh and Uttarakhand, it was able to solidify its control across the nation and the states. This gave these government leaders even more power to execute their plans.

In addition to its religious implications, the judgment delivered by justices Rajiv Sharma and Alok Singh in *Mohd. Salim v. State of Uttarakhand* and others was concerned with issues of federalism and tested whether a state, through its judiciary, could order the central government to take steps to protect the river (Ahmad 2017). The states have significant power over water in the state list of the Constitution. The high court directed that encroachers be evicted, and the central government was ordered to clarify the division of land and authority between Uttarakhand and Uttar Pradesh (Uttarakhand was carved out of Uttar Pradesh in 2000). But a tangle of contradictions was just beneath the surface. In the December 2016 hearing, the court ordered that there be no more delay in constituting a Ganga management board. This step had been mandated under section 80(2)(b) of the Uttar Pradesh Reorganization Act of 2000. After the reorganization into two states, this step was not done; as a result, there were faulty practices of control and maintenance along many canals in the system. The Ganga management board was supposed to assign jurisdiction of Ganga matters to the Namami Gange project at the central government level. The board was to include one representative from each of the states of Uttar Pradesh and Uttarakhand (but not the other three states in the basin). Two full-time members were supposed to be nominated by the state government as members, and the chairman was to be nominated by the central government. At the hearing in March 2016, an official from the Ministry of Water Resources told the court that despite the long correspondence, neither Uttar Pradesh nor Uttarakhand had cooperated with the central government in constituting the Ganga management board. This problem arose becauseof noncooperating state governments run by opposition parties to the ruling coalition in the central government. In the March hearing, the judges asked for finalization of committee membership within 60 days.

This situation is an odd reversal of assertions, with the state judiciary ordering the central government to constitute control over state water. Why would the high court justices do that? A potentially powerful Hindutva or Hindu nationalist message could be generated from the endeavor, and it would align with the previous call by local and national religious leaders to declare personhood for the Ganga. Moreover, environmentalists would celebrate the order. However, there is more.

This order draws a measure of legitimacy from India's legal philosophy seeking protection for the environment. Articles 48A and 51A(g) of India's Constitution refer to the duty of the state and the fundamental duty of an Indian citizen to protect and improve the environment and to take care of wildlife. Article 48A is often cited when passing judgments in favor of environmental protection (Ahmad 2017; Alley 2009). The invocation of fundamental rights in the protection of the environment has been extensive. In the current case on rivers, however, the immediate and more powerful administrative interest in the central government was to constitute the Ganga management board. By doing so, the central government could use a centralized method of planning for all water uses including hydropower, irrigation, potable supply, and wastewater and water treatment. The court stated, "The Constitution of the Ganga Management Board is necessary for the purpose of irrigation, rural and urban water supply, hydropower generation, navigation, industries. There is utmost expediency to give legal status as a living person/legal entity to rivers Ganga and Yamuna r/w Articles 48-A and 51A(g) of the Constitution of India."[19] In the March hearing, while fussing at all parties for delaying the constitution of the board, the judges issued the order on personhood for these rivers. The high court chastised the central government for dragging its feet on the very initiative it had moved the court to hear several months earlier.[20] This personhood rul-

---

19. Order dated March 20, 2017, in Writ Petition (PIL) No.126 of 2014, *Salim v. State of Uttarakhand*, para. 18.

20. The delay in the high court case also reflects the general operations of water informality that Alley (2016) outlined elsewhere and that other authors have explained in detail (Follmann 2014; Ranganathan 2016; Roy and Ong 2012; Schwartz et al. 2015). The central government, through the testimony of an officer in the Ministry of Water Resources, first demanded the constitution of the Board but then delayed announcement of membership, appearing to avoid compliance with the high court order. This flip-flop is, in fact, an operation of water informality whereby the state uses

ing became news across environmental policy communities as an achievement, but it ended up solidifying central government control over water, contradicting the jurisdictions embedded in the Constitution.

By the time elections were over in late March 2017, the two states of Uttarakhand and Uttar Pradesh had flipped control from the opposition party to the BJP. Now holding majority power at the state and central government levels, the goals are more aligned. It appears that the central government is not in the same rush to constitute the Ganga management board; center–state alignment might be achievable through the everyday working of overlapping bureaucracies. In contrast, the court was eager to enforce its earlier order.

Government Leaders, Legal Leaders, and Religious Values: Reproducing the Status Quo

In July 2017, the state government of Uttarakhand joined with the central government to submit a special leave petition (SLP) in the Supreme Court. With this SLP, they applied to stay and overrule the high court order on personhood. An SLP is an appeal from a lower court to the Supreme Court. The order came from a bench of Chief Justice of India J.S. Khehar and Justice D.Y. Chandrachud. The SLP was argued by a powerful Delhi-based lawyer in the capacity of barrister, which means he was hired for his symbolic power but not required to know or carry the case forward. The advocate on record, in theory, holds that responsibility. The use of this particular barrister meant that both state and central governments were serious about halting the ruling immediately and completely. After hearing preliminary arguments, the Supreme Court put a stay on the high court order stating, "The order had put the state government in a quandary. Since the rivers flow through several states, only the Centre could frame rules for their management. The ruling also raised questions like whether the victim of a flood in the rivers can sue the state for damages and also about whether the state and its officers will be liable in case of pollution in the rivers in another state through which it flows" (Express News Service 2017).

In this SLP, both the state and central governments appear interested in avoiding liability and responsibility for the grievances that people would

---

laws and principles that are appropriate for the moment but that may change those uses and even work contrary to them later.

be able to bring to the court in the name of Ganga's rights. However, more needs to be said about the fault lines within the legal apparatus itself, among the benches of the Supreme Court, the high courts, and the tribunal.

First, there is a need to look back to the NGT and understand its central problem—lack of independence. The central government has a role to play in the appointment, removal, and resignation of NGT judges, judicial members, and expert members.[21] Under section 10 of the NGT act, the central government can remove the chairperson and judicial and expert members in consultation with the chief justice of India. The Ministry of Environment and Forests also provides financial support to the tribunal. The central government exerts influence and control over many key aspects, which is important because many environmental cases are brought against the government. There are also cases in which the government supports an activity that causes environmental degradation, and the NGT is unable to take strong action in the matter, as the AOL case showed.

The rules of the NGT also allow bureaucrats holding government positions to be appointed to the tribunal. The Supreme Court struck down this practice in 2010 in *R. Gandhi*. In *R. Gandhi*, the Court held that a bureaucrat could hold a "lien" over his job for one year, after which time he would have to choose to one and leave the other. However, the rule makers of the NGT chose to ignore this case. In September 2015, the High Court of Delhi admitted a plea challenging the rules for appointment of the chairperson and judicial and expert members of the NGT. After retirement, there is also the possibility that judges and members can be absorbed into positions in various commissions. The former Chief Justice of India, Justice P. Sathasivam, for example, was appointed governor of Kerala by the central

---

21. As per the 2012 NGT rules, the selection committee is to consist of a chairperson who is a sitting or retired judge of the Supreme Court, appointed by the Chief Justice of India in consultation with the Minister for Law and Justice. In addition, an expert in environmental policy and an expert in forest policy are to be nominated by the Minister for Environment and Forests. The member secretary is secretary to the Government of India in the Ministry of Environment and Forests. The remaining two members are the chairperson of the tribunal and the director of the Indian Institute of Technology. The government has a role in appointing three members of the six-member committee, whereas one of the members is an acting official of the Ministry of Environment and Forests. Section 6 of the NGT act regarding the appointment of chairperson, judicial members, and expert members states that "1) the Chairperson, Judicial Members and Expert Members of the Tribunal shall be appointed by the Central Government; 2) The Chairperson shall be appointed by the Central Government in consultation with the Chief Justice of India; 3) The Judicial Members and Expert Members of the Tribunal shall be appointed on the recommendations of such Selection Committee and in such manner as may be prescribed" (see the National Green Tribunal Act of 2010, Act No. 19 of 2010, sec. 6).

government. Having an interest in future government appointments could affect the kinds of decisions judges take. The excessive influence of the central government and the interest of judicial members to maintain relations with the government for the sake of future appointments could provide another explanation for the contradictions that emerge in these three cases and in other cases where parties have political backing.

Finally, one cannot deny the benefit of having high courts hear environmental cases in view of the fact that there are only five benches of the NGT. Many applicants have to travel long distances for each hearing. This issue was raised in Parliament during discussion before the passing of the NGT act. Several parliamentarians were of the opinion that by taking over the powers of lower courts, the NGT would remove access to local justice from more than thirteen thousand districts and subordinate courts. This would be particularly difficult for economically weak members of the population.[22] Both the tribunal and the high court want a position of supremacy—the former as the designated forum for environmental justice and the latter as a constitutionally formed body. If an environmental court had been set up as originally suggested by the Supreme Court and the Law Commission, there would have been a forum with the legitimate power and requisite independence to adjudicate on environmental matters and deliver justice.

The tribunal can be credited with filling the gap in environmental governance and requiring, under the threat of punishment, some coordination among the government units in charge of pollution prevention. This role adds a level of third-party monitoring that is critically needed and crosscuts the contradictions between government responsibility and inaction. Nevertheless, because the NGT is far from independent, judges may use delay to appear to follow the law as they avoid the harshest punishments for powerful religious leaders.

Contradictions and Fault Lines

The conflicts surrounding these three cases show how fault lines exist over control of river water, uses of floodplains, and institutional controls. The first case shows that government leaders colluded with religious leaders

---

22. The Kanha-Pench case, however, highlighted the lack of clarity that exists regarding the hierarchy of the NGT and the high courts.

against environmental laws and policies to allow a religious leader to use the river floodplain and vacant government land. Government leaders allowed the organization's transgressions because the ruling party grows its symbolic capital from the public values of rivers as goddesses. The NGT chairperson and bench, although voicing condemnation of the religious organization, delayed decisions to negotiate the pressures and interests before them, manipulating temporality to mediate or balance the opposition between the citizen petitioner and the religious organization. The petitioner persisted amid administrative delay while religious values played a role in supporting the exemptions assumed by the religious organization.

The second and third cases show that fault lines exist across the configuration of governmental and legal institutions (see Studley [2018] on philosophical approaches to the juristic personhood of nature in different cases from around the world and Alley [2019] for a more detailed discussion of references on personhood and rights of nature). The high court asserted a new conservation formula for the sacred rivers that would have extended and redefined fundamental rights. However, the SLP carried the power of the central government in concert with the state government and blocked a redefinition of those rights and any new assertions on the large-scale controls over water. If the river's (or nature's) rights had been achieved, the river as deity would be entitled to own itself and the land and property surrounding it with a trust to administer the river's assets. This entity would have asserted a philosophical refiguring, reimagining Ganga and Yamuna not as all-powerful but rather vulnerable goddesses (Drew 2017; Haberman 2006). This form of ontological biopolitics is also glaringly contradictory—amid neglect of the river as a goddess by human administrations, there is a shifting call for her protection through a narrowing definition of her guardians. Moreover, her likeness to a woman and a person made that guardianship more urgent.[23] The rivers need to be saved, and one way to save them is to liken them to persons so that specific authorities, such as guardians, can take care of them or punish those causing her harm. Uma Bharti conveyed this idea in her speech to the climate conference when she likened Ganga's abuse to that suffered by a woman. Beyond Hindu ontologies, the actual way these rights (and care) would be asserted was not laid out in the initial

---

23. Foucault's distinction between this classical paradigm and the identification of a distinctively modern "biopolitics" is recalled, in which biological life (of both the individual and the species) becomes what is at stake in politics (for a history of this concept, see Foucault 2004; Lemke 2011).

ruling. How would a river assume her rights and get representation or even represent herself in arguing for her rights? How would a river, as goddess, sue or be sued?

In the AOL case and the cases on rights of nature, the rivers Yamuna and Ganga serve the interests of religious organizations and reinstate religious identities as they propose rivers' rights.[24] As the AOL case shows, however, the court can cave to a religious leader using the tactic of delay couched in legal and scientific rationalizations. The river was valuable for what it reaped the religious community and its leaders, as a commodity and as symbolic capital. While doing little to help solve a serious water problem, this religious leader could forge ahead because the community was committed to the event and the public interest could produce support for the ruling government and political party in the long run. Such commodification of river water cannot help solve critical pollution problems.

This discussion of river water conflicts leads to the planetary question of how these contradictions could frustrate resource decisions related to climate change. One way to understand the potential problems of climate change is to consider the hydrologic changes taking place as pollution or wastewater mounts in these surface waters. One of the planetary boundaries set by scientists when identifying the safe realm for life on earth is blue water consumption, along with recent modifications to the planetary definition (Gleeson et al. 2020). In the planetary boundaries framework, boundaries are set to propose where the fault line divides safe from unsafe life on the planet. This boundary is understood further as the point at which water resources can tip into a nonsustainable state for humans, other species, and the earth system. Humans are integral agents of the earth system, and their intensive and aggregated uses of fertilizers and fossil fuels put more nitrogen, phosphorus, and carbon dioxide into the atmosphere and onto the land. These activities change the climate system. The overproduction of nitrogen and phosphorus from the use of fertilizers and industrial

---

24. These cases contrast with other rulings in which the Supreme Court ruled in favor of expanding religious freedom. Berti (2017) notes, "We may refer here to the work of Ronojoy Sen [and Baxi] (2007) entitled 'Legalizing Religion' where he distinguishes between the American model of secularism which would be assimilative and the Indian model which is ameliorative. Sen notes how the state in India has pushed through its reformist agenda at the expense of religious freedom and neutrality and how the Supreme Court in India, particularly by using the 'essential practices doctrine', has contributed to a rationalization of religion and religious practices" (28; see also Berti [2015] on the judicialization of nature and Acevedo [2018] and Williams [2019] on the essential practices doctrine).

products leads to contamination of water bodies and changes the nitrogen cycle in ways that may produce harmful system feedbacks.

Carbon dioxide concentrations in the atmosphere block long-wave radiation bouncing off the earth, and the carbon dioxide molecules absorb this energy and create a warming effect. This warming triggers the intensification of the hydrologic cycle. This intensification brings more intense storms and precipitation events and affects the circulation of the Indian monsoon. Circulations shift over time, pushing previously wet regions into decadal drought. River pollution plays a strong role in limiting the viability of surface waters for humans, nonhumans, and ecosystems and reduces the consumption of blue water worldwide.

In this chapter, we have tried to understand why contradictions emerge in conflicts regarding river water uses and how they are tied to deeper fault lines of control. Central government trumps state governments in water conflicts, and religious interests can supersede legal controls sought by citizen pleas. Legal controls can also abort creative conceptual ideas, such as nature's right, and help government departments retain a primary grip on decisions that involve millions of cusecs or gallons of water. These case conflicts have crosscut surface and deep contradictions to maintain stability and the status quo. The reproduction of the status quo will continue to endanger water conditions and climate cycles in the near future.

References

Acevedo, Deepa Das. 2018. "Pause for Thought: Supreme Court's Verdict on Sabarimala." *Economic and Political Weekly* 53 (43). https://www.epw.in/journal/2018/43/commentary/pause-thought.html.

Ahmad, Omar. 2017. "A Court Naming Ganga and Yamuna as Legal Entities Could Invite a River of Problems." *Scroll.in*, April 3, 2017. https://scroll.in/article/833069/a-court-naming-ganga-and-yamuna-as-legal-entities-could-invite-a-river-of-problems.

Alley, Kelly D. 2002. *On the Banks of the Ganga: When Wastewater Meets a Sacred River.* Ann Arbor: University of Michigan Press.

———. 2009. "Legal Activism and River Pollution in India." *Georgetown International Environmental Law Review* 21 (793–819).

———. 2014. "The Developments, Policies, and Assessments of Hydropower in the Ganga River Basin." In *Our National River Ganga: Lifeline of Millions*, edited by Rashmi Sanghi, 285–305. Cham, Switz.: Springer.

———. 2016. "Rejuvenating Ganga: Challenges and Opportunities in Institutions, Technologies and Governance." *Tekton* 3 (1): 8–23.

———. 2019. "River Goddesses, Personhood and Rights of Nature: Implications for Spiritual Ecology." *Religions* 10 (9): 502. https://doi.org/10.3390/rel10090502.

Alley, Kelly D., Jennifer Barr, and Tarini Mehta. 2018. "Infrastructure Disarray in the Clean Ganga and Clean India Campaigns." *WIREs Water* 5:e1310.

Alley, Kelly D., and Georgina Drew. 2012. "Ganga." In *Hinduism*, Oxford Bibliographies, edited by Alf Hiltebeitel. Oxford: Oxford University Press. Accessed June 19, 2021. https://www.oxfordbibliographies.com/view/document/obo-9780195399318/obo -9780195399318-0034.xml.

Amirante, Domenico. 2012. "Environmental Courts in Comparative Perspective: Preliminary Reflections on the National Green Tribunal of India." *Pace Environmental Law Review* 29 (2): 441–69.

Ani. 2016. "Demonetisation: Sri Sri Ravi Shankar for Digital Financial Transactions in Rural Areas." *Business Standard India*, November 26, 2016. Accessed June 19, 2021. http://www.business-standard.com/article/news-ani/demonetisation-sri-sri-ravi -shankar-for-digital-financial-transactions-in-rural-areas-116112600456_1.html.

AOL (Art of Living Foundation). 2021. "About Us." Accessed June 19, 2021. https://www .artofliving.org/in-en/about-us.

Berti, Daniela. 2015. "Gods' Rights vs Hydroelectric Projects: Environmental Conflicts and the Judicialization of Nature in India." *Rivista Degli Studi Orientali*, 88:111–40.

———. 2017. "Animal Sacrifice under Judicial Scrutiny: Moral Reforms and Religious Freedom." Presented at "Taking Nature to the Courtroom" symposium, June 16, 2017, University of Edinburgh.

Chapple, Christopher Key, and Mary Evelyn Tucker. 2000. *Hinduism and Ecology: The Intersection of Earth, Sky, and Water.* Cambridge, MA: Harvard University Press.

Chezhiyan, Elan. 2008. "Examining the Need for Green Benches." Indlaw.com.

Donham, Donald L. 1999. *History, Power, Ideology: Central Issues in Marxism and Anthropology.* 2nd ed. Berkeley: University of California Press.

Drew, Georgina. 2017. *River Dialogues: Hindu Faith and the Political Ecology of Dams on the Sacred Ganga.* Tucson: University of Arizona Press.

Eck, Diana L. 1982a. *Banaras, City of Light.* New York: Knopf.

———. 1982b. "Ganga: The Goddess in Hindu Sacred Geography." In *The Divine Consort: Radha and the Goddesses of India*, edited by John Stratton Hawley and Donna Marie Wulff, 166–83. Boston: Beacon.

Express News Service. 2017. "SC Stays Uttarakhand HC Order on Ganga, Yamuna Living Entity Status." *Indian Express*, July 8, 2017. https://indianexpress.com/article/india /sc-stays-uttarakhand-hc-order-on-ganga-yamuna-living-entity-status-4740884/.

Financial Express. 2016. "Religious Leaders Submit Proposal to Government to Protect Ganga." *Financial Express*, September 2, 2016. https://www.google.com/amp/www .financialexpress.com/india-news/religious-leaders-submit-proposal-to-government -to-protect-ganga/365741/lite/.

Follmann, Alexander. 2014. "Urban Mega-projects for a 'World-Class' Riverfront—The Interplay of Informality, Flexibility, and Exceptionality along the Yamuna in Delhi, India." *Habitat International* 45 (3): 213–22.

Foucault, Michel. 2004. *The Birth of Biopolitics: Lectures at the Collège de France, 1978–1979* (Lectures at the College de France). New York: Palgrave Macmillan.

Gleeson, Tom, Lan Wang-Erlandsson, Samuel C. Zipper, Miina Porkka, Fernando Jaramillo, Dieter Gerten, Ingo Fetzer et al. 2020. "The Water Planetary Boundary: Interrogation

and Revision." *One Earth: Perspective* 2 (3): 223–34. doi:https://doi.org/10.1016/j
.oneear.2020.02.009.

Gopalan, Radha. 2017. "Why the Court Ruling to Humanise the Ganga and Yamuna
Rivers Rings Hollow." *The Wire*, March 27. Accessed June 17, 2021. https://thewire.in
/environment/ganga-yamuna-whanganui-human.

Haberman, David L. 2006. *River of Love in an Age of Pollution: The Yamuna River of Northern
India*. Berkeley: University of California Press.

Iyer, Aditya. 2017. "Maha Shivratri: Protesters Urge Modi Not to Attend Isha Foundation
Event." *Hindustan Times*, February 24, 2017. Accessed June 17, 2021. https://www
.hindustantimes.com/india-news/maha-shivratri-protesters-urge-modi-not-to-attend
-isha-foundation-event/story-LR9F1JjnIwE9uQ5je7gkJP.html.

Jain, Pankaj. 2011. *Dharma and Ecology of Hindu Communities: Sustenance and
Sustainability*. New York: Ashgate.

Kumar, Anuj. 2016. National Green Tribunal: A New Mandate towards Protection of
Environment." Accessed June 17, 2021. http://www.legaldesire.com/national-green
-tribunal-a-new-mandate-towards-protection-of-environment/.

Kumar, Bidisha, and George James. 2013. "Yamuna." In *Brill's Encyclopedia of Hinduism
Online*, edited by Knut A. Jacobsen, Helene Basu, Angelika Malinar, and Vasudha
Narayanan. Accessed June 17, 2021. https://referenceworks.brillonline.com/browse
/brill-s-encyclopedia-of-hinduism.

Lakshmana, K. V. 2017. "Still Waiting for Hearing on Isha Foundation's Illegal Construction
in TN: Activists." *Hindustan Times*, August 30, 2017. Accessed June 19, 2021. http://
www.hindustantimes.com/india-news/still-waiting-for-hearing-on-isha-foundation
-s-illegal-construction-in-tn-activists/story-8cKto2sgi65XVRWEVZxtVP.html.

Lemke, Thomas, ed. 2011. *Biopolitics: An Advanced Introduction*. New York: New York
University Press.

Lokgariwar, Chicu. 2017. "The Sad State of These Persons Called Ganga and Yamuna—
Can State Protect Them?" *Blog of the South Asian Network for Dams, Rivers
and People*, April 11, 2017. Accessed June 19, 2021. https://sandrp.wordpress
.com/2017/04/11/the-sad-state-of-these-persons-called-ganga-yamuna
-can-state-protect-them/.

Majumdar, Ushinor. 2016. "'I'll Take 5–10 Yrs to Restore the Floodplains': Ushinor Majumdar
Interviews C. R. Babu." *Outlook India*, June 3, 2016. https://magazine.outlookindia
.com/story/ill-take-5-10-yrs-to-restore-the-floodplains/297262.

Markandya, Anil, and Maddipati Narsimha Murty. 2000. *Cleaning Up the Ganges: A Cost
Benefit Analysis of the Ganga Action Plan*. Delhi: Oxford University Press.

Marx, Karl. (1939) 1973. *Grundrisse: Foundations of the Critique of Political Economy*. 6th ed.
Translated by Martin Nicolaus. New York: Vintage.

Nair, Harish. 2017. "Supreme Court Passes Buck to NGT over Cleaning Up River
Ganga." *India Today*, January 25, 2017. Accessed June 19, 2021. https://www
.indiatoday.in/mail-today/story/supreme-court-ngt-river-ganga-cji-khekar-modi
-government-956941-2017-01-25.

Nelson, Lance E. 1998. "The Dualism of Nondualism: Advaita Vedānta and the Irrelevance
of Nature." In *Purifying the Earthly Body of God*, edited by Lance E. Nelson, 61–88.
Albany, NY: SUNY Press.

Purkayastha, Shorbori. 2017. "Environmentalists Protest Ahead of Modi's Visit to Isha Foundation." *Quint*, February 24, 2017. Accessed June 17, 2021. https://www.thequint .com/news/environment/environmentalists-protest-modi-unveiling-112-feet-shiva -statue.

Ramasubramanian, R. 2017. "Citing Green Violations, Former Judge Urges Modi to Cancel Visit to Godman's Extravaganza." *The Wire*, February 23, 2017.

Ranganathan, M. 2016. "Rethinking Urban Water (In)formality." In *Oxford Handbooks Online*, edited by Ken Conca and Erika Weinthal. Oxford: Oxford University Press. doi:10.1093/oxfordhb/9780199335084.013.23.

Rauta, Rama. 2015. "The Ganga: A Lament and a Plea." In *Living Rivers, Dying Rivers: A Quest through India*, edited by Ramaswamy Iyer, 44–51. New Delhi, India: Oxford University Press.

Roy, A., and A. Ong, eds. 2012. *Worlding Cities: Asian Experiments and the Art of Being Global*. West Sussex, UK: Wiley-Blackwell.

Schwartz, Klaas, Mireia Tutusaus Luque, Maria Rusca, and Rhodante Ahlers. 2015. "(In) formality: The Meshwork of Water Service Provisioning." *WIREs Water* 2:31–36.

Sen, Ronojoy, and Upendra Baxi. 2007. *Legalizing Religion: The Indian Supreme Court and Secularism*. Washington, DC: East-West Center Washington.

Shivshankar, Goutham. 2017. "The Personhood of Nature." *Law and Other Things* (blog). http://lawandotherthings.com/2017/04/the-personhood-of-nature/.

Shrotria, Sudha. 2015. "Environmental Justice: Is the National Green Tribunal of India Effective?" *Environmental Law Review* 17 (3): 169–88.

Studley, John. 2018. "Juristic Personhood for Sacred Natural Sites: A Potential Means for Protecting Nature." *Parks* 24 (1): 81–96.

Tare, Vinod, and Gautam Roy. 2015. "The Ganga: A Trickle of Hope." In *Living Rivers, Dying Rivers: A Quest through India*, edited by Ramaswamy Iyer, 52–76. New Delhi, India: Oxford University Press.

Thakur, Joydeep. 2017. "NGT, Not Art of Living, Should Be Fined for Yamuna Floodplain Damage: Sri Sri." *Hindustan Times*, April 19, 2017. https://www.hindustantimes .com/delhi-news/sri-sri-says-ngt-not-art-of-living-should-be-fined-for-yamuna -floodplain-damage/story-SrCC3u9CexazwsnCTaQBSL.html.

Times of India. 2017. "SC Asks NGT to Monitor Steps to Clean Yamuna." *Times of India*, April 25, 2017. http://timesofindia.indiatimes.com/city/delhi/sc-asks-ngt-to-monitor -steps-to-clean-yamuna/articleshow/58351052.cms?from=mdr.

Van Buitenen, J. A. B. 2021. "Ramanuja." *Encyclopedia Britannia*. Accessed March 24, 2021. https://www.britannica.com/biography/Ramanuja.

Wainwright, William. 2006. "Concepts of God." In *The Stanford Encyclopedia of Philosophy*, edited by Edward N. Zalta. Updated February 1, 2017. https://stanford.library.sydney .edu.au/entries/concepts-god/.

Wilde, Lawrence. 1991. "Logic: Dialectic and Contradiction." In *The Cambridge Companion to Marx*, edited by Terrell Carver, 275–95. Cambridge, UK: Cambridge University Press.

Wilkes, Tommy. 2016. "Indian Guru's Festival on Delhi Floodplain Riles Greens, Worries Police." *Reuters*, March 10. http://in.reuters.com/article/us-india-environment -idINKCN0WB0EG.

Williams, Coleman D. 2019. *Freedom of Religion and the Indian Supreme Court: The Religious Denomination and Essential Practices Tests*. MA thesis, University of Hawai'i at Manoa.

KELLY D. ALLEY, PhD, University of Wisconsin–Madison, is Alma Holladay Professor Emerita of Anthropology at Auburn University. She is the author of book chapters and articles on religion, ecology, and environmental law and justice in India. Her book, *On the Banks of the Ganga: When Wastewater Meets a Sacred River* (University of Michigan Press 2002), explores Hindu interpretations of the sacred river Ganga in the context of current environmental problems. She has directed several research and outreach activities, most recently a National Science Foundation project on wastewater reuse in India. From this project, she wrote a book manuscript titled *Machines, Digestions and Wastewater Reuse in Contemporary India*. She has worked with numerous interdisciplinary teams and, in 2007, helped produce a radio series on the Ganga (Ganges) river for National Public Radio.

TARINI MEHTA is Associate Professor of Environmental Law and Assistant Dean of Student Affairs at Jindal School of Environment and Sustainability, O. P. Jindal Global University, India. She is also Director of the Environmental Law and Science Advocacy Forum, a multidisciplinary research forum that works to bridge the law-science gap through innovative programs and research. International and regional environmental and human rights law are her primary areas of research. She has a PhD in environmental law from Pace University, an LLM from the University of Cambridge, an LLB from the University of Warwick, and a BA in philosophy from the University of Delhi.

7

SUBVERSIVE COSMOPOLITICS IN THE ANTHROPOCENE

On Sentient Landscapes and the Ethical Imperative in Northern Peru

Ana Mariella Bacigalupo

How do people in La Libertad in northern Peru combine hope for the future—a vision of a better world that is beyond human control—with a kind of practical engagement with a world ravaged by climate change and the consequences of extractivism? The coastal Moche and Chicama valleys of La Libertad compose an area inhabited with "sentient landscapes" that have the capacity to sense and feel and the ability to act upon people. The people who live here are struggling to get by and to protect their health and surroundings in the context of mining expansion. We find *wak'as* (sentient landscapes made by indigenous ancestors) and *apus* (honorable indigenous ancestor mountains), as well as *curanderos* (shamans) and others, working to create a better future across species and landscapes. Those who live here call themselves "poor mestizos" and, in some sense, are liminal—in between—because they have both indigenous and Spanish backgrounds and reach across the realms of scientific and indigenous knowledge. Focusing on the mestizo response is crucial for understanding the complex practical moral reasoning provoked by floods and mudslides linked to the actions of sentient landscapes and the nature–culture relationships at the core of perceptions of environmental degradation in northern Peru. I analyze how these poor mestizos revive the leadership of indigenous sentient landscapes in ways that challenge the cultural legitimacy and moral authority of the Peruvian nation-state and subvert middle-class aspirations of global modernity (Bacigalupo 2018). Michael Herzfeld (2019) labels this process "subversive archaism."

On June 10, 2018, Percy Valladares Huamanchumo, a fifty-five-year-old mestizo with both indigenous Chimu and Spanish ancestry, fed the sacrificial rock at the base of apu Campana with Maltin Power, a soft drink made with malt. Apu Campana is the oldest and most powerful sentient mountain in the coastal Moche and Chicama valleys within La Libertad. In the past, Moche (200–900 CE) and Chimu (900–1470 CE) divine rulers fed Campana human sacrifices and other offerings in an attempt to control water resources, to prevent flooding caused by El Niño, and to promote environmental stability and communal health (Glowacki and Malpass 2003). Today, poor mestizos see wak'as and apus connecting the living with their indigenous ancestors, who control access to water, life, health, and fertility, as well as floods, mudslides, illnesses, and death.

Percy filled orifices in the sacrificial rock with water he had extracted from a well to appease its thirst in the desert environment. I lodged a mandarin and a piece of sweet sesame bread in crevices where previous visitors had left glass beads, fragments of painted Moche ceramics, seeds, shells, and powerful stones. "Water, liquids, and blood are the most precious offerings in the desert because they are life," Percy explained. "See how the rock is absorbing the Maltin. That's because Campana is an apu, a powerful ancestor and spiritual leader of all of us who live in the Moche and Chicama Valley. We feed the apu with offerings, and he gives us health, underground water to fill our wells, and rain that fills the rivers that come down from the Andes and allow us to irrigate our fields. But when people become corrupt and destroy the environment, the apus send floods and mudslides to punish them, because they need to be fed by humans in order to survive." Percy pointed to the place where the flood of 2017 had eaten away at the ritual site and carried away one of the stones: "This last flood was immense because the human corruption is immense. We need the guidance of the apus as moral leaders in this catastrophe."

Apus challenge Western assumptions about the separation of matter and spirit, and they speak to current attempts to decenter the human in what is known as the nonhuman turn. They communicate through the actions of animals and plants, through dreams, and through visions induced by Huachuma, the hallucinogenic San Pedro cactus and sentient being who opens the gates to other forms of consciousness (Sharon 2000; Glass-Coffin 2010).

A few weeks earlier, two local mestizo shamans described to me how they had channeled the higher consciousness of apus and wak'as in a ceremony in which we ingested a brew made from Huachuma, the hallucinogenic San Pedro cactus, enabling us to move beyond the boundaries of the human and connect with other beings. The stones, ceramic fragments, shells, plants, and sacred waters from different mountains on a curandero's ritual *mesa* (Gálvez 2014) enable these mestizo shamans to think like apus in much in the same way that the Amazonian shamans Eduardo Kohn (2013) works with are able to think like forests.

As in other places in Latin America (Kohn 2013; Gose 1994; Nash 1993; Taussig 1980; Salas 2019), sentient landscapes in northern Peru are self-absorbed beings when contacted individually. But for poor mestizos with whom I have worked, some sentient landscapes become superior moral persons and intentional actors in the struggle to counteract environmental damage since 2014 and 2017, when the Moche and Chicama valleys were devastated by the effects of mining and agribusiness, including floods and mudslides caused by torrential rains associated with El Niño but exacerbated by climate change. Poor mestizos attribute these floods to angry but moral landscapes forcing people to focus on the collective good by punishing those who are destroying the world and those who have failed to feed the apus and wak'as with offerings. Mestizo curanderos and their communities thus attribute a broader moral significance to environmental disasters and associate this destruction with climate change, systemic violence, and human health.

Poor mestizos raise questions about why some sentient landscapes rather than others acquire moral agency and how people might work with these landscapes to promote collective ethics and environmental and economic justice, despite the contradictions brought about by extractivism within their communities. Although many poor mestizos recognize mountains as ancestors with sentience and agency, their survival often depends on their ability to work in agriculture or other industries in which the exploitation of such mountains mirrors the exploitation of poor workers.

Gaston Gordillo (2021) has argued that people attribute human emotions to mountains in an attempt to understand their nonhuman power, which is both unknowable and highly disputed. But the views of curanderos are influential because they channel the mountains themselves.

Leoncio Carrión, known as Omballec, "Guardian of the Water," who controls the subterranean waters of the mountains, channels the thoughts of the apu Cuculicote:

> Everything is alive, even the stones and mountains vibrate. Our ancestors are part of the apu, the water, the wind. Cuculicote's hard rock face is heavy. His fangs hang out because he is angry at the stupidity of humans who destroy the world with mining for a coin. Sometimes he destroys the greedy world. At other times he heals [people] so that they develop a higher consciousness. . . . We need to develop a higher consciousness that is in tune with that of the apu. People have to learn to let go of their selfishness, be aware, and think more collectively with their communities, the earth, and the apus. (Interview, August 4, 2016)

Curandero Omar Ñique elaborates:

> The apus and wak'as are sending huge rains, floods, and mudslides to punish humans for our corruption and destruction of the earth's riches. We are so arrogant that we have lost the ability to understand that apus and wak'as have a higher consciousness and are the protectors of the resources of the Pachamama [Mother Earth]. Scientists explain the phenomenon of El Niño and climate change on television from a scientific perspective. But they do not understand that these are caused by the anger [reaction] of the apus and the wak'as, due to human corruption. But we know this, and that is why we are working together with our apus and wak'as to counteract this destruction and corruption through rituals and collective political actions through our environmental organizations. (Interview, August 26, 2016)

As moral and environmental crises wreak havoc around the world and extractive industries feed nondemocratic hierarchies, socioeconomic inequality, and violence, poor communities and activists are beginning to recognize the agency of sentient landscapes in mitigating these crises. Scholars have argued for the collapse of distinctions between nonhuman species, landscapes, and persons in their attempts to understand how environmental consciousness shapes alternative models of agency (Povinelli 2016; Haraway 2016; Descola 2013; Latour 1993; Ingold 2000). Others, following Isabelle Stengers (2010), have focused on how "cosmopolitics" (the ability of landscapes to act upon people) challenges Western perceptions of personhood and the sovereignty of culture over nature (Cruikshank 2005; Stengers 2010; Blaser 2013; de la Cadena 2015). Many indigenous Latin Americans have adopted transactional cosmopolitics (Kohn 2013; Gose 1994; Nash 1993; Taussig 1980; Salas 2019), meaning that they feed sentient landscapes with offerings to receive sustenance and control over environmental and economic resources—but the North Coast of Peru is unique. Why do some Peruvian sentient landscapes acquire moral agency? How do people engage with these landscapes to promote collective ethics while addressing the moral and environmental crises of a neoliberal world?

Academics have rarely thought about sentient landscapes in terms of environmental justice, collective ethics, and interethnic communities. Divisions among scholars have made this conversation difficult. Although the ontological approach focuses on interactions between indigenous human worlds and the nonhuman, including multispecies relationships and engagements with sentient landscapes, the political ecological approach reckons with human destruction of the earth.

Scholars using an ontological approach have characterized sentient indigenous landscapes as existing in a radically different world incommensurable with modern politics and alien to the lives of mestizos (de la Cadena 2015, 275; Kohn 2013). Sentient places may participate in community rituals and protests (de la Cadena 2015, 96–97, 134), but they remain outside modern political issues such as negotiations with the state to recover community places and participation in place-based, ethnic environmental movements (de la Cadena 2015; Viveiros de Castro 1998; Kohn 2013; Taussig 1987). Marisol de la Cadena (2015) writes, "To save the mountains from being swallowed up by mining corporations, activists themselves—*runakuna* included—withdrew earth beings from the negotiation. Their radical difference exceeded modern politics, which could not tolerate there being anything other than a cultural belief" (275). But these interpretations do not take into account the predicament of indigenous life in the modern world (Ramos 2012) or the perceptions and practices of poor mestizos as inhabitants of a realm neither indigenous nor elite. Kohn (2013) studies the politics of sentient forests within indigenous Amazonian environmental activism, but mestizos remain marginal to his analyses.

Sentient landscapes have not been considered political subjects in place-based, ethnic environmental movements, such as those studied by Arturo Escobar (2016). Scholars have made sophisticated analyses of the relations between indigenous myths and rituals and modern states (Hill 1988; Santos Granero 1998) but have not explored sentient places as persons participating in interethnic politics.

Some interpret the personhood of sentient landscapes as "romanticized and naïve" because this concept challenges modernist ideas of reality

(Fabricant 2017).[1] Others argue that sentient places are not "real" because indigenous people doubt their existence or because they do not appear in "traditional" narratives (Cepek 2016). These views ignore typical contradictions between discourse and practice, the role of joking and context (Willerslev and Pedersen 2010), and the ways in which people constantly reinvent "the traditional." And despite their interest in values and moral selves, scholars have been less concerned with a collective ethics that promotes political change. These scholars have not gone further to imagine how sentient places might be understood by persons participating in interethnic negotiations or how the invocation of sentient landscapes can lead to environmental justice (Hornborg 2017, 2).

Neither a radical ontology nor a political ecology approach is useful for understanding how poor mestizos work with wak'as and apus for environmental and political ends. Both approaches obscure the role played by sentient places in environmental activism and modern politics, although for different reasons. Furthermore, the conflicts between scholars endorsing these different theories have little bearing on the actual practices of poor mestizos. Some anthropologists in South America negotiate a middle ground between these conservation and extractive economies (High and Oakley 2020), with their radically different understandings of reality. I am interested in exploring a new understanding of cosmopolitics where sentient landscapes become agents of social justice and reconciling these different positions from the ground up. Poor mestizos in La Libertad give us just that type of pragmatic and experiential perspective.

I am interested in sentient landscapes as agents of social justice, and I describe how poor coastal mestizos appropriate the moral agency of these landscapes for social and environmental transformation. By defining *community* and *well-being* as humans in relationship to places as persons, poor mestizos resignify "nature" itself as an anchor for social justice and collective ethics. They believe that incorporating apus and wak'as into their local environmental movements offers the only viable strategy for counteracting the power of extractive industries and the environmental devastation they produce.

My analysis is based on five summers (2015–19) of collaborative ethnographic research with communities in the coastal Moche and Chicama

---

1. Nicole Fabricant, comments on panel, "New Landscapes of Extractivism in the Andean Region," American Anthropological Association Meetings, November 30, 2017, Washington, DC.

valleys and the Andean area of Huamachuco in La Libertad, Peru. During this time, I explored the practical dimensions of moral reasoning by participating in a variety of ritual, everyday, and political contexts with mestizo people: shamans, fishermen, patients, miners, priests, professionals, and activists. I visited apus and wak'as, tended medicinal plant gardens, and helped with rituals, fishing and mining, and community events. I worked closely with fifteen curanderos (mestizo shamans) and five guardians of specific apus and wak'as to understand how they come to identify with and engage them through prayers, offerings, and healing and divination rituals. I documented the kinds of relationships wak'as maintain with other wak'as, mines, and curanderos; with the plants, animals, and rocks that inhabit these places; and with weather and the greater environment. I asked curanderos and others how they interpret the emotions, power, status, and agency of different wak'as and apus. I have interviewed them about their apocalyptic dreams and visions of wak'as and recorded their observations about the effects of climate change, which they use to predict floods and mudslides. Finally, I observed poor mestizos and curanderos participating in environmental movements, seeing how they strengthen their relationships with sentient places, plants, animals, and communities and how they educate their communities about the relationship among wak'as, climate change, and morality. I have also visited the archives and museums of the area to study how mestizos and their engagements with sentient landscapes have changed over time.

The Historical and Ethnic Contexts of Mestizo Cosmopolitics

Who are these poor mestizos? And what shapes their identities, spirituality, healing practices, and engagement with sentient landscapes? The term *mestizo* is itself fluid and politically contested. Because *mestizaje* is neither a state policy nor an overarching form of identity in Peru, race is understood in cultural rather than biological terms and is tied to class, education, and ethnic discrimination among white, mestizo, and indigenous groups (De la Cadena 2000).

Poor mestizos on the northern coast complicate highland versus coastal divisions between race and class that shape engagement with apus and wak'as. Because the Spanish term *indio* (Indian) is an insult across Latin America, most indigenous people refer to themselves as *people* in their indigenous languages (*runa* means "person" in Quechua). The Spanish term *serrano* is used to refer to all Andean Indians. Racist Peruvian ideologies construe Quechua-speaking serranos as uneducated, irrational peasants living in kinship-based groups with territorial ties to specific sentient mountains (De la Cadena 2000) who play central roles in indigenous community rituals tied to Catholic saints (Mendoza 2006, 2017). And they construe serranos as standing in the way of educated, urban Spanish-speaking mestizos on the coast who do not engage sentient landscapes (Mendez 2011; de la Cadena 2000; Mendoza 2006; Gose 1994). "Race" and racialization are produced in habitual forms of social interaction between rural first-language-Quechua speakers and urban first-language-Spanish speakers (Huayhua 2014; Mannheim 2015). The Quechua serrano elite identify as mestizo and celebrate indigenous culture but discriminate against poor rural serranos (De la Cadena 2000), although, in practice, rural serranos also participate in the creation and celebration of indigenous culture (Mendoza 2006). Coastal mestizos are like the serrano elite in that they ascribe racialized attributes to Quechua peasant serranos, whom they view as inferior, but they also challenge this dichotomy.

Poor mestizos in La Libertad are monolingual Spanish speakers who do not identify as serranos but maintain historical ties with the northern highland communities in Huamachuco through intermarriage and economic, social, and ritual exchange. They foster belonging in their interethnic communities by incorporating into their communities poor serrano and jungle migrants with whom they work in mining, agribusiness, and sugarcane factories and with whom they practice cosmopolitics. "In Huanchaco," says Percy, "the wisdom on which local power is based has also been absorbing the cultures of people who come from different areas. And that's why it still exists."

Although poor mestizos in La Libertad distinguish themselves from the educated, upper-class, urban mestizos in Trujillo and Lima, they also borrow hegemonic nonindigenous concepts such as spirituality, sacred places, the environment, and cultural and natural patrimony, which distinguish between nature and culture, to make their cosmopolitics understandable to the larger public. But because mestizo cosmopolitics intersect with regional and national politics, migration from the highlands to the coast, and actively shared community places, it also disrupts divisive highland versus coastal racial and ethnic identity politics.[2]

---

2. I developed the term "mestizo cosmopolitics" in 2014 to refer to how poor mestizos in northern Peru engage sentient landscape ancestors through rituals and political events and the ability of these landscapes to act upon human lives, realities, and political practices. Sarah Bennison (2016) separately and concurrently developed the term mestizo cosmopolitics to refer to the practices of Andean Peruvians in Huarochiri in central Peru.

Coastal mestizos like Percy Valladares understand landscapes some-what differently than rural serranos in southern Peru. For Quechua ser-ranos, *apu* is an honorary title sometimes given to mountains thought to be sentient, whereas *wak'as* are attributes, events, and things with agency whose personhood is reproduced through social practice. Archaeologists have emphasized that both apus and wak'as, in this usage, are material entities separate from human engagement (Mannheim and Salas 2015).

Complex historical and social processes have enabled the emergence of a contemporary mestizo space-based spiritual and environmental politics. Indigenous Moche (200–900 CE) and Chimu (900–1470 CE) peoples were first colonized by Andean Incas (1470–1532), who promoted engagement with apu and wak'a persons in Quechua terms, then by Spanish (1532–1824) and Peruvian (1824–present) priests, who attempted to extirpate such "idol-atries," demonize shamans, and drive engagements with apus and wak'as underground. Many still reject apus and wak'as as a form of demonic ani-mism because they threaten Catholic monotheism, Western divisions of human and nonhuman, and the interests of extractive industries.

Nevertheless, apus and wak'as have reemerged publicly among poor mestizos with indigenous and Spanish ancestry now that *curanderismo* (a hybrid form of shamanism) is recognized as national patrimony and curanderos channel the thoughts of apus and wak'as through dreams and visions induced by ingestion of the San Pedro cactus. Curanderos draw their power from specific sentient landscapes, and environmental destruction associated with climate change is linked to systemic violence and human health.

Curanderos have tempered their colonial history and usurped the hier-archy of the Catholic Church by resignifying Saint Cyprian as an ambiva-lent curandero/apu and Jesus as a moral one. As curanderos participate in the spiritual tourism industry, the concepts of apu and wa'ka are acquir-ing global currency. The healing dimensions of curanderismo reemerged among coastal mestizos in the late 1980s after the discovery of tombs of Moche divine rulers and shamans and the rise of spiritual tourism. Ameri-can anthropologists and Peruvian academics legitimated the syncretic healing of curanderos through their writings (Joralemon and Sharon 1993; Glass-Coffin 1998 2010; Gálvez 2014) and collaborative conferences with curanderos (Morales 2012; Vasquez and Flores 2009). Their practice of place-based environmental and spiritual politics unsettles the dichotomies of Western thought and reflects their effects to engage sentient landscapes.

Although both populations from the coast and the Andes in La Libertad now identify as mestizo rather than indigenous and are monolingual Spanish speakers, their pattern of exchange remains strong (Schaeffer 2006). Both poor, monolingual Spanish-speaking serranos and poor coastal people in La Libertad are victimized by extractive industries, seen as obstacles to progress, and self-identify as poor mestizos. Rather than an instrumental "political exploitation of neo-Indianity" (Galinier and Molinié 2013, 189), poor mestizos' appropriation of coastal apus and wak'as challenges the serrano–coastal mestizo dichotomy. This happens by incorporating into their communities serranos from La Libertad with whom they work in mining, agribusiness, and sugarcane factories; by celebrating their coastal Moche and Chimu indigenous roots; and by embracing "irrational" relationships with coastal apus and wak'as on their own terms. In these ways, poor mestizos distinguish themselves from the educated upper-class urban mestizos in Trujillo and Lima and imagine themselves as part of a moral universe in which sentient landscapes are social beings who possess indigenous powers and participate in political activism. Because apus and wak'as work for the broad environmental, social, and political interests of poor mestizos, they open a new kind of political debate based on the value of a shared interethnic world.

Environmental Disasters and Extractivism in La Libertad

Extractive industries including archaeological excavations have produced conflicting interests in communities that also make offerings to apus and wak'as. Confrontations between mestizos and the government began in 2014, when the region was devastated by the effects of mining, agribusiness, and floods exacerbated by climate change. Curanderos and other poor mestizos in the valleys attribute floods to angry but moral landscapes forcing people to focus on the collective good by punishing those who are destroying the world. By collaborating with apus and wak'as, poor mestizos ground their collective ethics in place and grapple with resource exploitation, climate change, and greed as a single problem.

Throughout the previous decade, the Peruvian government had intensified resource extraction to boost economic growth, simultaneously accelerating greenhouse warming, deforestation of the Amazon, reduction of Andean glaciers, and pollution of water, air, and soil. Peru is now the third most vulnerable country in the world with regard to climate change

(DW 2018). Peruvian leaders and the Ministerio de Energía y Minas (Ministry of Mining and Energy) believe that reconstruction will solve the environmental problems; however, their treatment of the economy, environment, and human health as unconnected policy domains has had disastrous consequences. State-supported mining and sugarcane factories produce and reproduce nondemocratic hierarchies, structural violence, and environmental devastation; their exploitation of nature is mirrored in their exploitation of poor workers. Like gold miners in the Peruvian Amazon (Ulmer 2020), poor miners in the Andes and coast contend with the disposability of their life and labor. Poverty has increased in the valleys as mudslides and flooding have destroyed much-needed irrigation canals and as the crises of the sugarcane industry have led to lost jobs for mestizo laborers and lower pay for those still working (Kus 1989; Klaren 2005).

Marginalized mestizos' survival often depends on their ability to work in fishing, agriculture, or extractive industries, but the contamination of their air and water produces a variety of health problems—communicable diseases, respiratory illnesses, cancer, depression, alienation, and envy—and destroys the medicinal plants used by curanderos to heal them.

Andean Huamachuco is one of the poorest districts in the region of La Libertad. It has been devastated by gold-mining companies, which have coerced locals into selling their lands and water sources as the price of gold soars. International mining companies and informal, illegal mining have contaminated the area. Mining has contaminated the rivers that flow down through the Moche and Chicama valleys, which have also been contaminated by the sugarcane industries and coastal mining.

Health problems produced by environmental degradation disproportionately affect the poor. Weak infrastructure and discrimination also compromise poor mestizos' agency and hinder their ability to be diagnosed and treated (Bussmann and Sharon 2007). Corrupt officials pocket resources meant to help poor communities access health interventions and rebuild after floods and mudslides, leading to further crises.

The government speaks of resource nationalism and state-led development, but it is clear that foreign capital is in control: Canadian and American mining companies and Italian and German sugarcane factories reap economic benefits, ignoring immense social and environmental costs and the resistance from the people negatively affected. Extractivism exacerbates poverty in mestizo communities through dispossession, which encourages the accumulation of wealth elsewhere. Although neoliberalism has failed to improve economic growth and, in fact, has increased poverty, contemporary right-wing and left-wing Peruvian presidents have employed resource extraction as a primary development strategy. Neither side has been able to deliver on promises of greater social inclusion and economic justice. These failures have sparked political unrest in indigenous and mestizo peasant communities, where the harm engendered by the present order of accumulation and consumption is felt most intensely and where the contradictions of development are most visible (Hickel 2014).

Poor mestizos in La Libertad use various strategies to manage the effects of neoliberal policies. As seen in studies on the "environmentalism of the poor" (Martinez-Alier 2002), some contest the unequal distribution of ecological costs and benefits. Like antidevelopment scholars (Escobar 1995), poor mestizos question Western ideologies that set the terms for how the poor can live rather than allowing local communities to address their own problems. At the same time, mestizos acknowledge that they have less power and fewer resources with which to develop their communities because of the stratifying effects of capitalism. They criticize the economic and political strategies used by the state, extractive industries, and development agencies to limit mestizos' ability to develop. Some mestizos try to resolve this inequality by establishing relationships with mining and sugar companies in hopes of gaining enough wealth and social mobility to better their lives. Others are painfully aware that they have no mobility in this system and that the extractive industries will only undermine their heath, environment, and well-being.

Transactional Cosmopolitics between Mestizos and Wak'as

Like indigenous people in Latin America and Moche and Chimu divine rulers of the past, most poor mestizos perceive apus and wak'as as non-human persons who depend on the actions of humans for their survival. Sentient places are shaped by what humans make of them, and in Peru, they take part in a culturally specific practice of commensality—the sharing of food and resources. Mestizos do not conceive of apus and wak'as as what Elizabeth Povinelli (2016) calls "affective without intention," their sentience and power uncontrollable and irreducible to what humans make of them. Perception is never unmediated, and, as Bruce Mannheim and Guillermo Salas (2015) have shown, sentient places are built simultaneously from a culturally specific material practice of commensality and a universal theory of mind that is presupposed by commensality.

In the Andes, both the earth and emplaced wak'as are thought to have a fertile life-giving side that grants resources (referred to as Pachamama in some areas) and a destructive life-taking side (referred to as Supay or El Tío; Nash 1993 Taussig 1980). Humans engage in a transactional logic of exchange to reproduce the personhood of wak'as by feeding them with *pagos a la tierra* ("payments to the earth," also known as *despachos*), which include items such as coca leaves, corn, chicha, sweetbreads, llama fat, and fruit. In turn, wak'as grant humans permission to visit them, in addition to sustenance, gifts, health, and other resources (Mannhein and Salas 2015). Percy Valladares shared with me the consequences of his offering to Cerro Campana, the oldest and most powerful apu:

> I made an offering of coca to the wak'a and then my mind went blank and I re-appeared disoriented about 150 meters farther down the slope. I sat on a stone and looked down and saw a silver triangle, part of a pectoral, and I looked around and there were fragments of ceramics. I saw a human bone; as I went to bury it, I found offerings of shells, beads, quartz. . . . Then I cut my hand badly and I finally managed to stop the bleeding with a bandage, and as I came down there was a beautiful ceramic waiting for me, which wasn't there before. The wak'a received my offering [of blood] and then pulled me to this place to grant me gifts. (Interview, March 11, 2018)

If humans do not feed wak'as, recognize their personhood, and treat them with respect, or if they are overexploited, then wak'as become angry and cause torrential rains, floods, mudslides, and earthquakes and inflict illnesses, accidents, bad luck, and death.

Curanderos have relationships with specific wak'as, whom they invoke and feed to gain resources, to grant healing and well-being to their patients, or inflict illness, infertility, or death on their enemies. The energy of the wak'a changes according to how it is used:

> The energies that emanate from the wak'as can be positive or negative depending on who uses them and for what purpose. If humans conceive the wak'a as a place that emanates the love generated by humans and people respect the wak'a and give it offerings, it will be a place of healing, it will have good energy. If humans conceive the wak'a as a place where the hatred generated by humans is increased and people mistreat and exploit the wak'a, it will be a place of sorcery, it will have bad energy. Wak'a Cortada in El Brujo archaeological complex has traces of bad energy because people often use it for sorcery. I was doing ceremony there and I heard the sound of chains dragging on the

floor, people groaning, swords clashing, and a horse galloping and neighing in fear. These are the negative energies that remain there. (Omar Ñique, interview, August 1, 2017)

The interaction between humans and the life-giving wak'a is sometimes imagined as a reciprocal exchange involving respect for nature and the controlled use of its resources, which bolsters the power of the wak'a and the well-being of the community. A *pago a la tierra* for life-giving purposes is similar to the offering peasants make to agricultural lands for the purpose of enhancing fertility and crop yield (De la Cadena 2015). Lucila, a curandera from the highland community of Yamobamba, Huamachuco, explained how her community feeds the wak'a mountain Siempre Viva (Always Alive) and how it grants them resources, health, and well-being in exchange: "The first of May we go up to my mountain and celebrate it. We feed it *chuño* [dried potatoes], alcohol, sweetbreads, because we are there all night. We give it pig. We bury it in the earth. Then we do ceremonies to heal people and make them flourish, to allow them to do well on their path. The mountain gives us abundance, health, and happiness" (interview, August 18, 2016).

Exchanges between poor mestizos and wak'as in La Libertad today often relate less to the ideal of balanced reciprocity and more to the human desire for personal gain while avoiding punishment for a failure to make offerings, for contaminating wak'as, or for overharvesting. Alongside a wak'a logic, mestizos have internalized Catholic or Christian beliefs that humans are superior to nature. According to Catholicism, human exchanges with wak'as are a form of ecocentric paganism that undermines "the value of people over and above all other living things" (Catholic News Agency 2009). When wak'as punish mestizos for a lack of reciprocity, wak'as gain a reputation for being dangerous to humans. Catholic priests and saints often intervene to become protectors of poor mestizos against "diabolic" wak'as who make demands, kill humans, or cause illness. Mestizos' reaction to the wak'a Lake Sausachocha's punishment of humans by drowning and disappearing is a case in point. A family photographer explains:

Sausachocha has killed many disrespectful people. Some tourists came to shit on the banks, and the rain carried that waste into the water. They fished many trout, and Sausachocha slapped them around with waves and drowned them.... Since she is an *encanto* [charmed place], she swallowed several people and their animals on the banks. And since people complained, the Franciscans came and baptized Sausachocha and made her Catholic to try to calm her. . . . However, she still drowns disrespectful people. Some drunk musicians came to the pier and decided to swim. They made no offerings to Sausachocha. I told them not to go in because Sausachocha would swallow them and they would get hypothermia, but they told me to shut up and one of them began to sink and he drowned. (Interview, August 17, 2016)

These local cosmopolitics are further complicated as mestizos respond to the larger political and economic implications of extractivism. Extractivism partakes of both a neoliberal logic and an Andean logic related to the destroying or life-taking dimension of the earth (Nash 1993), which Michael Taussig (1980, 147) refers to as "the devil" and links to commodity fetishism. According to the neoliberal logic, humans exploit nonrenewable natural resources for financial gain without considering the consequences to wak'as, the environment, or the collective good of communities in the area. Mining, for example, disrupts the logic of reciprocal exchange because the process of extracting gold pollutes and destroys a wak'a in part or in its entirety.

Some mestizos are willing to destroy part of a wak'a through mining because they believe they and their families will become wealthy and socially mobile, although they rarely do. Large mining companies have the technology to transform environmental resources into money, whereas poor mestizos do not. Workers are paid low salaries by mining companies and lack the resources to mine informally. They often make *pagos a la tierra* to a mine so that it will yield gold and protect them from harm, but this rarely works. They suffer the health consequences of mining and are punished by wak'as for polluting them, and they fail to become wealthy. Many mestizo miners say that wak'as punish them for stealing the mountains' gold by making them poor, ill, or impotent and by causing accidents.

The case of Luis, a poor mestizo from highland Huamachuco, exemplifies this dynamic. Although Luis healed his family with medicinal plants and declared himself a "protector of nature and wak'as," he worked at El Toro mine and practiced informal, illegal mining. Illness and poverty led him to rent out his machines and turn to Catholicism to protect himself from the revenge of the wak'a he mined:

First, I worked for the mining company El Toro, which is the oldest mine here. Later, the Canadian mining companies came with a huge capacity to invest and took over. In those older times, we workers extracted gold with pickaxes. One day when I was hacking at the mountain, it screamed at me. I ran out of the mine. I became very ill. Then I found gold on my own land. I chewed coca leaves while I thought about the prospect of gold mining on my own land. The coca leaves were bitter, a bad sign. But I decided to exploit it anyway because I was poor.

I had to hire people to work it. But I had to pay each man fifty *soles* a day, and there were eight men so it was too much expense. I could not compete with the Canadians. Then my water became contaminated, and everything started going backward. I had bad luck. The mountain was angry with me and made me poor and ill. I stopped working and started renting my machines to extract minerals to the mining companies for 150 *soles* a day. That is how I live. I became sicker, and I had no energy left to work. I had trouble breathing. I cleaned the machines with Florida water and holy water at night so that Catholicism would protect me and the mountain would not kill me. (Interview, July 20, 2015)

Poor mestizos also work under the supervision of archaeologists who excavate wak'as, seen as pre-Columbian sites of scientific value. The excavations destroy parts of the wak'as, and locals argue that this destruction decreases their powers, although not to the same extent that mining and looting do. Archaeologists uncover structures and extract pre-Columbian artifacts and human bones to study them and place them in museums, while thieves extract objects to sell on the black market and make money. Because the artifacts do not go back to the community, some poor mestizos view archaeologists as *wak'eros profesionales* (professional looters) and try to stop them through sorcery.

Some mining companies and archaeologists embrace the wak'a logic and hire curanderos to make enormous *pagos a la tierra* and appease apus and wak'as so that they will yield gold or reveal their pre-Columbian treasures without killing anyone or making them ill. Curanderos Agustin Fernandez and Leoncio Carrión were hired to perform *pagos a la tierra* to the wak'a El Brujo. A mining company in Huamachuco also hired Omballec to perform rituals and prepare *pagos a la tierra*, which Omballec did so the mine would yield gold and not harm poor miners. "I gave the wak'a *pagos* and begged it to yield gold so that the owner will be richer," he said. "I was sad that the wak'a would be destroyed. I asked it not to punish the poor workers, because it was the owner of the mine who was responsible" (interview, August 10, 2016).

But despite *pagos a la tierra* and rituals at wak'as, sudden deaths and accidents remain pervasive at mining sites and archaeological digs in La Libertad and among *wak'eros*. There have been shootings at Chan-Chan, multiple fatal accidents on the Pan-American Highway at the foot of Cerro Campana, and a huge number of accidents and illnesses among workers and archaeologists at the excavations of wak'a La Luna and wak'a El Brujo. Archaeologist Enrique Zavaleta explains, "We have had many strange accidents at wak'a La Luna. . . . Not everyone makes *pagos a la tierra* . . . . Some workers feel sick or develop respiratory illnesses. One young man got *aire de wa'ka* and a curandero sprayed him with perfume and put cotton in his ears and *ruda* in his nose. Other have had *aire de muerto* [air of the dead], which is from the spores of fungi or the gas antimony, which is trapped in tombs where there have been metals" (interview, March 11, 2018).

Omballec argues that the extraction of a wak'a's riches and pre-Columbian artifacts is a theft of the wealth and life force of the wak'a, an evil act. This theft triggers a response from the life-taking side of the wak'a, who becomes hungry and must be fed with blood to compensate for the life force that has been taken. He explains, "If you want to do harm or steal water and riches from the wak'as and the apus, you have to sprinkle blood on the ground, and then the spirits with whom you have been communicating will come and eat to recover their life force. Our Moche and Chimu ancestors fed wak'as and apus with human sacrifices. . . . If you don't offer a sacrifice or sprinkle blood, then the spirit will start taking human lives by causing accidents, deaths, floods, and mudslides" (interview, August 9, 2016).

Poor mestizos' interactions with apus and wak'as and their life-giving and life-taking forces allow them to make sense of the of the structural violence—the poverty, illness, and death that rule their precarious lives—and to establish a matrix of human and nonhuman accountability around extraction. Ultimately, however, the sentience and power of landscapes lie beyond human control and not always in direct relation or response to human concerns—for example, when feeding an apu is followed by scarcity, illness, or flooding. Some mestizos respond by depersonalizing sentient landscapes and arguing that they are not sentient at all. Others try to make sense of these events by arguing, as Omballec does, that the punishing side of apus and wak'as emerges when they have been overexploited. These destructive forces have acquired moral implications in northern Peru—compelling people to work for the collective good by taking revenge on those who abuse the environment, on poor mestizos, and on the community.

Moral Cosmopolitics, Prophetic Critique, and Climate Change

Poor mestizos are living in a politically and morally fraught time in which the planetary changes caused by humans are inseparable from capitalist forms of governance and their exploitative patterns of extraction and accumulation. The end result of neoliberalism is environmental devastation, global climate change, and a society that no longer respects people and their values (Klein 2014; Castillo and Egginton 2016). Although the neoliberal vison and the values of curanderos, apus, and wak'as within mestizo communities come into conflict, it is the values of the latter that figure in the moral cosmopolitics of La Libertad. By working with apus and wak'as as politically active beings, poor mestizos ground their collective ethics in place and grapple with resource exploitation, climate change, and greed as a single problem.

After the devastation of their communities by mudslides and floods in 2014, poor mestizos in the Moche and Chicama valleys broadened the scope of their work with wak'as and apus. They began to speak of wak'as not as self-absorbed entities but rather as superior, moral persons who critique human corruption and greed and to include them as intentional actors in the struggle to counteract the environmental devastation caused by neoliberalism and climate change. Wak'as and apus punish humans with the effects of climate change—deluges, mudslides, warming waters—to force them to reach a higher consciousness that focuses on the collective good and the environment. Omballec told me, "In March 2017, the clouds and lightning accumulated on the mountain, and the water came down in waves. The apus, they threw the rain to cleanse themselves. . . . The water came down with such anger to teach us to respect nature, apu, wak'a. People came to ask me to calm the Cuculicote Mountain and tell it to stop. But this is not something that we should try and stop. Apu cannot be dominated. It is nature with superior powers telling humans to change their behavior and to develop a higher consciousness, to think collectively" (interview, August 11, 2016).

Mestizos draw on indigenous apus and wak'as not only because they play a central role in mestizos' identities and lives but also because sentient landscapes offer the only viable strategy for counteracting environmental devastation in the long run. Poor mestizos are attempting to challenge the power of extractive industries and the government by incorporating apus and wak'as into Asociación de Rescate y Defensa del Apu Campana (Association for the Rescue and Defense of Apu Campana) and the Colectivo Comunidad Consciente (Collective of Community Consciousness), two local environmental movements created by curanderos in 2014 in the fishing town of Huanchaco. The association defines Campana—the tallest mountain near Trujillo—as an apu: "Apu Campana is the highest spiritual and gnostic authority of our indigenous past who nurtures our collective and ecological values"; it is "central to our collective health as our healers invoke it and use its medicinal plants to heal" and is a "fundamental natural and cultural patrimony of the region."[3]

In July 2018, the association petitioned for the legal personhood of Apu Campana and denounced its destruction. Poor mestizos contest the unequal distribution of ecological costs and benefits and attempt to re-create a world of value by applying fundamental indigenous forces to political purposes.

The collective includes the participation of apus and wak'as as intentional agents and is led by curandero Omar and administrator Percy. They argue that capitalist frameworks imagine nature as inexhaustible and separate from human life, transforming reality itself into a commodity and individuals, corporations, and states into corrupt market agents who exploit the environment, a noncommodity that is indispensable to the survival of all.

Through these organizations, poor mestizos produce documents aimed at protecting and recovering lands from invaders, agribusiness, and mining companies. Percy explains:

> Wak'as and apus are our sacred ancestors, imbued with power, who give organizing principles for the lives of all human and non-human people and stress morals and respectful relationships between themselves, humans, the environment, and collective well-being. The Colectivo's goal is to use these principles to create a platform for collective decision-making on issues that have to do with regional environmental agendas, natural and cultural patrimony, and anticorruption campaigns in the management of disasters. We value the collective good, transparency, and democratic co-governance and are against corruption, exploitation, and abuse by extractive industries and the government. We seek to improve the environmental laws to protect green spaces and work with environmental organizations, people who defend natural and cultural patrimony, natural people, wak'as and apus, and authorities who are committed to the Colectivo's goals to create a better world. (Interview, February 21, 2018)

Through wak'as and apus, locals negotiate for spiritual protection from government officials, mining companies, and others. By incorporating

---

3. Escritura Pública de Constitución de Asociación Civil Denominada Asociación de Rescate y Defensa del Apu Campana, Santuario Ecológico y Arqueológico de la Libertad-4056-2012.

spiritual and moral factors into plans for growth and development, poor mestizos try to manage natural, social, and spiritual capital for the welfare of future generations. Percy Valladares told me, "The Colectivo has been involved in the protection [of], respect [for], and conservation of sacred spaces, such as Cerro Campana, Quebrada Santo Domingo, Cerro Cuculicote, Cerro La Virgen, Chan-Chan, [and] wak'a El Brujo. The Colectivo defends these places not only as structures, but also as entities with power and energy that flow from each wak'a and apu toward the inhabitants" (interview, February 21, 2018).

Curanderos like Omar and Omballec, in turn, engage wak'as for ritual healing purposes and as political actors in the collective. They use the collective to educate their communities about the connection between climate change, morality, floods, and mudslides, as well as the actions of wak'as. Omar explains, "I tell but also show people how the disasters they are suffering are caused by wak'as, who control climate change. And [if] they act responsibly toward all the environment and become moral citizens who fight to conserve wak'as and apus as natural and cultural patrimony and to oppose greed and corruption, they will be saving their own lives and the world" (interview, July 25, 2017). The collective draws on local curanderos, lawyers, journalists, and activists, who become the voices of local grassroots movements; they comprise real and virtual platforms, with an online journal focused on social responsibility.

Despite an interest in values (Robbins 2007; Keane 2015) and moral selves (Laidlaw 2002; Lambek 2010; Fassin 2012), scholars have not addressed an ethics that promotes political change. Those who study cosmopolitics have generally not discussed contemporary political activism involving sentient landscapes and not focused enough on how humans and nature influence each other or challenge a Western capitalist model in which individuals, communities, nations, and the globe are separate entities. In contrast, I argue that sentient landscapes are agents of social justice (Povinelli 2016). Poor coastal mestizos appropriate the moral agency of these landscapes for social and environmental transformation.

Activists, communities, and national governments have recognized the legal personhood of nature, which is an extraordinary conceptual and legal advance that allows for the prosecution of polluters under personal injury laws (Cano Pecharroman 2018). India, Australia, and New Zealand recognize the legal personhood of rivers; Bolivia and Ecuador recognize the personhood of Pachamama (Mother Earth); and Colombia recognizes the legal personhood of the Amazon rain forest. Laws have established the right of nature to live, to continue vital cycles and processes free from human alteration, to have pure water and clean air, and to not be polluted or genetically modified. The goals of these laws include radical new conservation and social measures to reduce pollution and increase the regulation of mining and other industries.

The Peruvian government recognizes corporations as persons but not the legal personhood of nature. On April 2, 2018, President Martín Vizcarra passed a law to strengthen climate policy, to prepare people for extreme natural events, to generate sustainable development projects, and to reduce greenhouse gas emissions. Nevertheless, the implementation and enforcement of such laws are difficult because they conflict with goals for national economic growth and the interests of extractive industries. Despite these impediments, in July 2018, the association petitioned for recognition of the legal personhood of apu Campana and denounced its destruction. At the time of this writing, the organization is working with local lawyers to modify laws on the rights of persons in Peru and to create new ones that recognize the legal personhood of nature; it is also addressing the barriers surrounding these legal initiatives. However, the moral dimension of sentient landscapes remains the focus.

Amid these political and legal difficulties, the everyday cosmopolitics of poor mestizos challenges academic distinctions between the real and the metaphorical in significant ways (Descola 2013; Viveiros de Castro 1998; Vilaça 2005; Ingold 2000; Kohn 2013). Poor mestizos may disavow the existence of apus and wak'as when they are interacting with state officials who assume a modernist, secular perspective, but they still engage with apus and wak'as to make sense of their everyday lives and environmental devastation. Locals often move pragmatically between these worlds according to their experiences, needs, and audiences, and some make spirits and sentient landscapes the objects of irony and joking (see Willerslev and Pedersen 2010). Pablo, a taxi driver from Trujillo, exemplifies this process:

> We have never had El Niño phenomena like this before. . . . There were seven *waycos* [mudslides]. You could say to the government, it is because the city was constructed in the path of the water. But human corruption is to blame. We laughed at nature. I didn't believe in the apus so that they could not punish me and then I could become rich working in

the mines. . . . We destroyed them. But the apus got us anyway. Now the apus are punishing us with climate change. My house was destroyed in the flood. We are being punished because of the lack of morality as everyone does things for themselves without thinking about community. Now that we lost our homes, we are learning to respect the apus again and join the Colectivo because this is also about our survival and our future. (Interview, July 24, 2017)

Mestizos speak of wak'as and apus as agents and also use them as tropes for morality or environmental justice, reviving indigenous notions of the personhood of nature as a strategy with which to gain support for political movements around environmental justice. Poor mestizos situate themselves simultaneously within sentient places and within modern politics. Locals' knowledge is always contextual (Briones 2014)—sometimes it refers to a different world, and sometimes it is immersed in modern discourses—but for poor Peruvian mestizos, the personhood of wak'as, inherent or ascribed, is never doubted.

Apocalyptic Predictions through Visions and Observations

Many poor mestizos believe that modern society's project of indefinite growth and progress is coming to an end, halted by finite resources. They predict a millenarian, cyclical Andean *pachacuti*, an overturning of space and time to return to a previous era and their rulers (Millones 2007). In this specific case, there would be an overturning of the space–time of modern industrial civilization and reestablishment of the rule of sentient landscapes in an indigenous world order. Christophe Bonneuil (2015) describes this ecocatastrophic narrative as one that calls for a radical change in dominant ways of living, consuming, and producing and that rejects technological means of saving the planet. "The apocalypse will affect the whole of humanity," says Omballec, "although the poor will suffer from the floods the most, even though the greed of the industries contributes most to the problem. But the moral poor who continue to make offerings will return to the new era when wak'as and apus will rule" (interview, July 25, 2019). The beginning of this *pachacuti* is marked by the rains and floods predicted by curanderos.

In the past, Peruvians mocked curanderos' predictions about El Niño phenomena, which were based on a local science they could not understand. Omballec explains:

In 1983 we had a drought, and a journalist came to ask for my predictions about El Niño. He published an article in the newspaper mocking me for saying it would rain in February and March. But then the rain started pelting down and all the ravines we visited became one huge river. [There were] huge mudslides. People lost their homes . . . and others began to use the water and mud [for] their crops. From 1998 onward, climate change and the destruction of wak'as and apus made El Niños worse, and people desperately wanted predictions. The archaeologists learned that I have my own science and come and ask me when it [is] going to rain to protect their sites. A scientist came to ask me, too, and since my prediction coincided with his, he published it. (Interview, July 27, 2017)

Curanderos meld the agency and morality of apus and wak'as expressed through floods, mudslides, tsunamis, dreams, and visions with scientific discourses and their own observations of rising temperatures and changes in the movements of currents, winds, and animals and in the blooming patterns of plants to make their knowledge accessible to a large interethnic audience. Because poor mestizos have seen the connections among the actions of apus and wak'as, curanderos' biometric readings, and environmental science, they have also understood the consequences of overexploitation simultaneously in environmental, moral, and social terms. "Climate change has changed the time for rains," Omballec told me. "Before, it rained in October or November, but now that has changed to January and February. And the last El Niños have been in March and April" (interview, August 9, 2016). He went on:

I knew that El Niño was coming in 1983, 1988, 2014, 2017, and at other times because the apus showed me through visions, and the animals and the plants showed me through their actions. The termites began to crawl out of the wood and the dragonflies would feed on them, so there were many dragonflies of all colors. The spiders, insects, and snakes came down from the mountain and into the streets and houses, seeking refuge so that they would not drown. There were a huge number of herons . . . and the birds and animals were alarmed. I lit a candle and it sputtered, a sign that there was a lot of human evil and immorality and the apu would punish that with water. The pacay and guaba trees bloomed early and lost their flowers and got new ones. The zapote tree, which is small and grows in the ravines, started growing really tall and getting a lot of green leaves on top because there was a lot of water accumulating underground. (Interview, July 20, 2017)

Biologist Carlos Quiroz commented, "Omballec is right. When the rain begins in the highlands, the first thing that fills up is the *phreatic* water layer underground, and once those are saturated, then the river flows above. Many small springs can also emerge, and the seeds that were underground sprout . . . and the whole desert then becomes green with running water, frogs, insects, deer, so it is clear that a big rain is coming" (interview, July 5, 2018).

Many curanderos, poor mestizos, and even mestizo scientists talk about climate change both as a message sent by apus in dreams and visions and as a scientific prediction. Quiroz combines scientific measurements with his own experiences with the apus:

> People dream and have visions about apus being angry because of environmental devastation and apus as causing earthquakes and exploding with water to destroy the world. These experiences have a scientific basis. As temperatures rise because of El Niño phenomena, the atmospheric pressure over the mountains falls and then wind moves in from areas of high pressure to areas of low pressure to compensate. The mountains of the coastal range are affected by the tropical rain, which saturates the aquifers under the rocky surface. The low pressure areas generated by the high temperature of the atmosphere make the water in the water tables rise and reach the surface of the mountains, dramatically increasing the water volume of the ravines.
>
> Earthquakes exacerbate this effect and therefore are often occur together because of the water pressure produced by rain infiltrated under the surface. Apus also show you things that science cannot. We took the San Pedro cactus brew with a curandero named Julio, and he saw the Apu Campana as a mountain of crystal. The next day, we found ancient Moche crystal mines that Moche artisans used to make marvelous necklace beads. (Interview, July 5, 2018)

One man from the area predicted enormous destruction in the region in the next five years as a result of subterranean waters flowing from the mountain apus. He articulated this prediction by combining scientific discourses with local narratives about apu ancestors:

> There are two mountains who will destroy this area in the next five years: San Ildefonso will empty its water on Trujillo and Chan-Chan, and the ravines from Campana will flood and destroy Huanchaco. San Ildefonso is the mother: she is receptive to all the mountains and has no tensions with any of them. Cuculicote is the healing mountain. San Ildefonso, the mother, loves her children, but she also teaches and punishes them. The Campana is the great-grandfather and the warrior, and it has tensions with all the mountains. The mountain is becoming really green because it is filling with water. . . . Before, the winds on Campana ran from north to south; now, they go from south to north. . . . The hot and cold air will clash and create cyclones. Campana

becomes drier while other mountains are green, filled with water. . . . The clash of the cold Humboldt current with the warm waters from El Niño will cause cyclones in the Pacific, and tsunamis of more than thirty meters [will] come from the sea [and] flood the area all the way to Chocope and Ascope. (Interview, July 28, 2017)

Or, as Omballec said almost a year earlier, "Humans will eat disasters and anguish today and tomorrow" (interview, August 28, 2016).

By promoting limits on growth, working to reverse the destruction caused by modern industrial civilization, and arguing that we need to change our lives and the global order, poor mestizos create the possibility of a new era of participatory politics in which moral humans and sentient landscapes create collective ethics and community life together.

Conclusion

Both matter and spirit, the subversive sentient landscapes of northern Peru have much to contribute to current discussions about the social and environmental inequalities produced by the global economic order. Although spirituality and sentient landscapes are often considered marginal to these discussions, I have shown that they play a central role in challenging Western values and practices that exploit people and the environment in devastating ways.

A study of the cosmopolitics of landscapes with sentience and agency offers us news ways of thinking about the Anthropocene—the epoch of human influence on the earth—while providing a model for radical environmental–political action. Scholars such as Andreas Malm and Alf Hornberg (2014) have critiqued the Anthropocene as a depoliticizing concept that neglects political economy and histories of capitalism because it attributes environmental degradation to all humankind rather than to a specific sector of the world's population. Others, such as Heather Swanson (2016) and Dipesh Chakrabarty (2009), argue that the concept can be engaged in diverse ways and that such studies can politicize colonial histories and human inequalities and lead to a radical politics. I have focused on the radical politics enabled by the resistance of local populations to environmental changes. My work with communities practicing an indigenous, place-based environmental and spiritual politics asks how imperial capitalism and its effects can be questioned to engage nature as a subjective being at local and global scales. I argue that the vulnerabilities produced by climate change have catalyzed a rethinking of values and the development of movements for change that reconcile the cosmopolitics of sentient places with scientific study and political activism.

This work goes beyond the limitations of approaches that focus only on different ways of being or only on the relationship between political and environmental issues. Poor mestizos use sentient landscapes to mobilize based not on the divisive identity politics of indigenous versus nonindigenous peoples but rather on the sense of a shared interethnic world. Their appropriation of the powers of sentient landscapes moves us beyond academic distinctions between indigenous and nonindigenous, human and nonhuman, nature and culture, ontology and political activism. Instead, through ritual and political engagements with apus and wak'as, they construct a better world in which corrupt and immoral humans and angry but moral wak'as and apus are together accountable for climate change, which locals describe simultaneously in scientific terms, through local observations, and from dreams and visions. The morality and agency of apus and wak'as and the actions of curanderos and communities together shape the well-being of people and the planet.

Acknowledgments

Funding for this research was provided by the generous support of the Wenner Gren Foundation, the Fulbright Foundation, and the OVPRED/ HI Seed Money in the Arts and Humanities fund, a Community for Global Health Equity Grant, and a Civic Engagement Grant at the State University of New York–Buffalo. I presented versions of this chapter at several seminars and colloquia, and I am grateful to the participants for their comments. Many thanks to the participants in the Yale Macmillan Center colloquium in agrarian studies organized by James C. Scott; the seminar at the Center for the Study of World Religions at Harvard Divinity School, organized by Charles Stang; the workshop "Engaging Religion" at the Ansari Institute for Global Engagement with Religion, University of Notre Dame, organized by Thomas Tweed; the seminar of the Latin American Studies Program, University of Toronto, organized by Susan Antebi; the anthropology colloquium in the Department of Anthropology, Boston University, organized by Merry White; the workshop "Religion and Climate Change in Regional Perspective" at the Center for Latin American and Latino Studies at American University, organized by Evan Berry; and the participants on the panel, "Subversive Agencies: Sacred Landscapes and Climate Change in the Anthropocene," November 14–18, 2018, organized by Robert Albro and Ana Mariella Bacigalupo.

References

Bacigalupo, Ana Mariella. 2018. "'La Política Subversiva de los Lugares Sentientes': Cambio Climático, Ética Colectiva y Justicia Ambiental en el Norte de Perú." *Scripta Ethnologica*, 49:9–38.

Bennison, Sarah. 2016. "Who Are the Children of Pariacaca? Exploring Identity through Narratives of Water and Landscape in Huarochirí, Peru." PhD diss., Newcastle University.

Blaser, Mario. 2013. "Ontological Conflicts and the Stories of Peoples in Spite of Europe: Toward a Conversation on Political Ontology." *Current Anthropology* 54 (5): 547–68.

Bonneuil, Chrisophe. 2015. "The Geological Turn: Narratives of the Anthropocene." In *The Anthropolocene and the Global Environmental Crisis: Rethinking Modernity in a New Epoch*, edited by C. Hamilton, F. Gemene, and C. Bonneuil, 15–31. London: Routledge.

Briones, Claudia. 2014. "Navegando creativamente los mares del disenso para hacer otros compromisos epistemológicos y ontológicos." *Cuadernos de Antropología Social* 40:49–70.

Bussmann, Rainer, and Douglas. Sharon. 2007. *Plants of the Four Winds: The Magic and Medicinal Flora of Peru*. Trujillo: Graficart.

Cano Pecharroman, Lidia. 2018. "Rights of Nature: Rivers That Can Stand in Court." *Resources* 7 (1): 13. doi:10.3390/resources7010013.

Castillo, David., and William. Egginton. 2016. *Medialogies: Reading Reality in the Age of Inflationary Media*. New York: Bloomsbury.

Catholic News Agency. 2009. "Nature Must Not Be Valued Above Man, Pope Warns." Catholic News Agency, December 15, 2009. Accessed June 21, 2021. https://www.catholicnewsagency.com/news/18058/nature-must-not-be-valued-above-man-pope-warns.

Cepek, Michael. L. 2016. "There Might Be Blood: Oil, Humility, and the Cosmopolitics of a Cofán Petro-Being." *American Ethnologist* 43 (4): 623–35.

Chakrabarty, Dipesh. 2009. "The Climate of History: Four Theses." *Critical Inquiry* 35 (2): 197–222.

Cruikshank, Julie 2005. *Do Glaciers Listen? Local Knowledge, Colonial Encounters, and Social Imagination*. Seattle: University of Washington Press.

De la Cadena, Marisol. 2000. *Indigenous Mestizos: The Politics of Race and Culture in Cuzco, Peru, 1919–1991*. Durham, NC: Duke University Press.

———. 2015. *Earth Beings: Ecologies of Practice across Andean Worlds*. Durham, NC: Duke University Press.

Descola, Phillippe. 2013. *Beyond Nature and Culture*. Chicago: University of Chicago Press.

DW. 2018. "Perú, pionero en Latinoamérica con su Ley Marco de Cambio Climático." Deutsche Welle (DW), April 17, 2018. Accessed June 21, 2021. https://www.dw.com/es/per%C3%BA-pionero-en-latinoam%C3%A9rica-con-su-ley-marco-de-cambio-clim%C3%A1tico/a-43426965.

Escobar, Arturo. 1995. *Encountering Development: The Making and Unmaking of the Third World*. Princeton, NJ: Princeton University Press.

———. 2016. *Territorios de diferencia: Lugar, movimientos, vida, redes*. Popayán: Editorial Universidad del Cauca.

Fassin, Didier., ed. 2012. *A Companion to Moral Anthropology*. Malden, MA: Wiley-Blackwell.

Galinier, Jacques., and Antoinette Molinié. 2013. *The Neo-Indians: A Religion for the Third Millennium*. Boulder: University Press of Colorado.

Gálvez Mora, Cesar. 2014. Una mesa de curandero y la geografía sagrada del Valle de Chicama. En Por la mano del hombre. Prácticas y creencias sobre chamanismo y curandería en México y el Perú. Lima, Peru: Asamblea Nacional de Rectores.

Glass-Coffin, Bonnie. 1998. *The Gift of Life: Female Spirituality and Healing in Northern Peru*. Albuquerque: University of New Mexico Press.

———. 2010. "Shamanism and San Pedro through Time: Some Notes on the Archaeology, History, and Continued Use of an Entheogen in Northern Peru." *Anthropology of Consciousness* 21 (1): 58–82.

Glowacki, Mary, and Michael Malpass. 2003. "Water, Huacas, and Ancestor Worship: Traces of a Sacred Wari Landscape." *Latin American Antiquity* 14 (4): 431–48.

Gordillo, Gaston. 2021. The Power of Terrain: The Affective Materiality of Planet Earth in the Age of Revolution." *Dialogues in Human Geography* 11 (2): 190–94. https://doi .org/10.1177/20438206211001023.

Gose, Peter. 1994. *Deathly Waters and Hungry Mountains: Agrarian Ritual and Class Formation in an Andean Town*. Toronto, Can.: University of Toronto Press.

Haraway, Donna. 2016. *Staying with the Trouble*: Making Kin in the Chthulucene. Durham, NC: Duke University Press.

Herzfeld , Michael. 2019. "What Is Polity? Subversive Archaism and the Bureaucratic Nation-State 2018 Lewis H Morgan Lecture." *HAU: Journal of Ethnographic Theory* 9 (1): 23–35.

Hickel, Jason. 2014. "Book Review: *The New Extractivism: A Post-neoliberal Development Model or Imperialism of the Twenty-First Century?* Edited by Henry Veltmeyer and James Petras." *LSE Review of Books* (blog), October 20, 2014. Accessed September 2, 2021. https://blogs.lse.ac.uk/lsereviewofbooks/2014/10/20/book-review-the-new -extractivism-a-post-neoliberal-development-model-or-imperialism-of-the-twenty -first-century-edited-by-henry-veltmeyer-and-james-petras/.

High, Casey, and Elliott Oakley. 2020. "Conserving and Extracting Nature: Environmental Politics and Livelihoods in the New 'Middle Grounds' Amazonia." *Journal of Latin American and Caribbean Anthropology* 25:236–47.

Hill, Jonathan. 1988. *Rethinking History and Myth: Indigenous South American Perspectives on the Past*. Urbana: University of Illinois Press.

Hornborg, Alf. 2017 "Convictions, Beliefs, and the Suspension of Disbelief: On the Insidious Logic of Neoliberalism." *HAU: Journal of Ethnographic Theory* 7 (1): 553–58.

Huayhua, Margarita. 2014. "Racism and Social Interaction in a Southern Peruvian combi." *Ethnic and Racial Studies* 37 (13): 2399–417.

Ingold, Tim. 2000. *The Perception of The Environment: Essays on Livelihood, Dwelling and Skill*. London: Routledge.

Joralemon, Don, and Douglas. Sharon. 1993. *Sorcery and Shamanism: Curanderos and Clients in Northern Peru*. Salt Lake City: University of Utah Press.

Keane, Webb. 2015. *Ethical Life: Its Natural and Social Histories*. Princeton, NJ: Princeton University Press.

Klaren, Peter. 2005. "The Sugar Industry in Peru." *Revista de Indias* 65 (233): 33–48.

Klein, Naomi. 2014. *This Changes Everything: Capitalism vs the Climate*. New York: Simon & Schuster.

Kohn, Eduardo. 2013. *How Forests Think: Toward an Anthropology beyond the Human*. Berkeley: University of California Press.

Kus, James S. 1989. "The Sugar Cane Industry of the Chicama Valley, Peru." *Revista Geográfica* 109:57–71.

Laidlaw, James. 2002. "For an Anthropology of Ethics and Freedom." *Journal of the Royal Anthropological Institute* 8:311–32.

Lambek, Michael, ed. 2010. *Ordinary Ethics: Anthropology, Language, and Action*. New York: Fordham University Press.

Latour, Bruno. 1993. *We Have Never Been Modern*. Cambridge, MA: Harvard University Press.

Malm, Andreas, and Alf Hornberg. 2014. "The Geology of Mankind? A Critique of the Anthropocene Narrative." *Anthropocene Review* 1 (1): 62–69.

Mannheim, Bruce. 2015. "All Translation Is Radical Translation." In *Translating Worlds, The Epistemological Space of Translation*, edited by Carlo Severi and William F. Hanks, 199–219. Chicago: University of Chicago Press.

Mannheim, Bruce, and Guillermo Salas. 2015. "*Wak'as*: Entifications of the Andean Sacred." In *The Archaeology of Wak'as: Explorations of the Sacred in the Pre-Columbian Andes*, edited by T. Bray, 47–72. Boulder: University Press of Colorado.

Martínez-Alier, Joan. 2002. *The Environmentalism of the Poor: A Study of Ecological Conflicts and Valuation*. Cheltenham, UK: Edward Elgar.

Mendez, Cecilia. 2011. "De Indio a Serrano: Nociones de raza y geografía en el Perú, siglos XVIII–XXI." *Histórica* 35 (1): 53–103.

Mendoza, Zoila. 2006. *Crear y Sentir lo Nuestro: Folklore Identidad Regional y Nacional en Cuzco, Siglo XX*. Lima, Peru: Pontificia Universidad Católica del Peru.

———. 2017. "The Musical Walk to Qoyllor Rit'i: The Senses and the Concept of Forgiveness in Cuzco, Peru." *Latin American Music Review* 38 (2): 128–49.

Millones, Luis. 2007. "Mesianismo En América Hispana: El Taki Onqoy." *Memoria Americana* 15:7–39.

Morales, Ricardo. 2012. "Curanderos y académicos: una experiencia en Trujillo (1994 y 1995)." *Pueblo y Continente* 23 (1): 14–18.

Nash, June. 1993. *We Eat the Mines and the Mines Eat Us: Dependency and Exploitation in Bolivian Tin Mines*. New York: Columbia University Press.

Povinelli, Elizabeth. 2016. *Geontologies: A Requiem to Late Liberalism*. Durham, NC: Duke University Press.

Ramos, Alcida. 2012. "The Politics of Perspectivism." *Annual Review of Anthropology* 41:481–94.

Robbins, Joel. 2007. "Between Reproduction and Freedom: Morality, Value, and Radical Cultural Change." *Ethnos* 72:293–314.

Salas, Guillermo. 2019. *Lugares parientes: Comida, Cohabitacion y Mundos Andinos*. Lima: Editorial Universidad Católica del Perú.

Santos Granero, Fernando. 1998. "Writing History into the Landscape: Place, Myth, and Ritual in Contemporary Amazonia." *American Ethnologist* 25 (2): 128–48.

Schaeffer, Nancy Ellen. 2006. "Fishermen, Farmers, and Fiestas: Continuity in Ritual of Traditional Villages on the Northwest Coast of Peru." PhD diss., University of Texas, Austin.

Segura Vásquez, Nyler. 2009. "Curanderismo del complejo cultural costa norte: un itinerario para reflexión." In *Revista Medicina Tradicional. Conocimiento milenario*, serie Antropología 1, Trujillo, Museo de Arqueología, Antropología e Historia.

Sharon, Douglas. 2000. *Shamanism and the Sacred Cactus*. San Diego Museum Papers 37. San Diego, CA: San Diego Museum of Man.

Stengers, Isabel 2010. *Cosmopolitics I (PostHumanities)*. Translated by Robert Bononno. Minneapolis: University of Minnesota Press.

Swanson, Heather. 2016. "Anthropocene as Political Geology: Current Debates over How to Tell Time." *Journal Science as Culture* 25 (1): 157–63.

Taussig, Michael. 1980. *The Devil and Commodity Fetishism in South America*. Chapel Hill: University of North Carolina Press.

———. 1987. *Shamanism, Colonialism, and the Wild Man: A Study in Terror and Healing*. Chicago: University of Chicago Press.

Ulmer, Gordon. 2020 "The Earth is Hungry: Amerindian Worlds and the Perils of Gold Mining in the Peruvian Amazon." *Journal of Latin American and Caribbean Anthropology* 25:324–39.

Vilaça, Aparecida. 2005. "Chronically Unstable Bodies: Reflections on Amazonian Corporalities." *Journal of the Royal Anthropological Institute* 11:445–64.

Viveiros de Castro, Eduardo. 1998. "Cosmological Deixis and Amerindian Perspectivism." *Journal of the Royal Anthropological Institute* 4:469–88.

Willerslev, Rana, and Morten Pedersen. 2010. "Proportional Holism: Joking the Cosmos into the Right Shape in North Asia." In *Experiments in Holism: Theory and Practice in Contemporary Anthropology*, edited by Ton Otto and Nils Bubandt, 262–78. Malden, MA: Blackwell.

ANA MARIELLA BACIGALUPO is Professor of Anthropology at the State University of New York–Buffalo. She is the author of *Thunder Shaman: Making History with Mapuche Spirits in Patagonia* (University of Texas Press, 2016); *Shamans of the Foye Tree: Gender, Power, and Healing among the Chilean Mapuche* (University of Texas Press, 2007); *The Voice of the Drum in Modernity: Tradition and Change in the Practice of Seven Mapuche Shamans* (Universidad Católica de Chile Press, 2001); and *Hybridity in Mapuche "Traditional" Healing Methods: The Practice of Contemporary Mapuche Shamans* (PAESMI, 1996). She also coauthored *Modernization and Wisdom in Mapuche Land* (San Pablo Press, 1995).

8

CLIMATE VULNERABILITY AS THEOLOGICAL BRIDGE CONCEPT

Examples from Puerto Rico

Andrew R. H. Thompson

On September 20, 2017, Hurricane Maria hit Puerto Rico as a category four hurricane with maximum sustained winds of 155 miles per hour. For months afterward, Puerto Rico reeled and staggered in its recovery. Electricity was lost across the territory for weeks and, in many places, several months. Food, water, fuel, and health care remained scarce months later. Roads were blocked, agriculture was decimated, and houses were demolished. Competing accounts circulated regarding the number of fatalities linked to the hurricane, ranging from an official government count of sixty-four to hundreds to more than one thousand (Schwartz 2018; Robles 2017; Hernández 2018).

Puerto Rico is a clear illustration of climate vulnerability. Its fate is the likely fate of any number of island and coastal regions in the not-too-distant future, as climate change is predicted to increase the magnitude and frequency of extreme weather events (and is already doing so). More specifically, the hurricane's impact on Puerto Rico illustrates the multifaceted nature of climate vulnerability. The territory's geophysical vulnerability was compounded by multiple other vulnerabilities. It is economically impoverished, with high unemployment, crumbling infrastructure, and widespread lack of access to many goods and services. Its status as an unincorporated territory of the United States fractures its political autonomy. All of these factors increase Puerto Rico's sensitivity to extreme weather events and undermine its ability to recover from them (its resilience). Political, economic, social, and psychological dynamics combine with geophysical characteristics and increased exposure to climatic events to increase overall vulnerability.

In this chapter, I will explore these factors of vulnerability to suggest that vulnerability is a multifaceted concept that can help link disparate dimensions of the conversation about climate change in ways that have not yet been fully appreciated. Specifically, I will argue that vulnerability can be useful as a theological bridge concept: based on the resonances between a scientific or policy-oriented account of vulnerability and a theological one, the theological account can interrogate and expand the scientific one, pointing out unexpected features and possibilities. To make this argument, I will examine accounts of vulnerability presented in policy discourse, including assessment reports by the Intergovernmental Panel on Climate Change (IPCC) and a variety of scholarly studies. I will then consider the use of the term *vulnerability* in theological accounts, especially as it is explicated in the work of Sarah Coakley (2002) and Elizabeth O'Donnell Gandolfo (2015). This theological examination will indicate areas of overlap between the policy and theological accounts that point to unexpected strengths present in the vulnerability concept. Finally, I will consider some practices, both religious and secular, in climate-vulnerable islands and island nations, especially Puerto Rico,[1] that reflect this power in vulnerability.

I note here, and will reiterate at the conclusion, that this argument is not meant to suggest that vulnerability is good in itself, to enjoin vulnerability on already vulnerable communities, to blame the vulnerable for their vulnerability, or to imply that vulnerable communities ought to accept or even be grateful for their vulnerability. As the theological account will make clear, although some forms of vulnerability are inherent to human life and even to flourishing, other forms are violations of human good and thus are to be resisted. That account will claim that the seeds of such resistance are present in vulnerability itself, and part of my goal is to build on this claim to suggest some ways that communities might cultivate those seeds. More broadly, though, my purpose is to commend the theological account to those interested in understanding climate vulnerability—this is

---

1. For the sake of accuracy, it is worth pointing out that Puerto Rico is neither an island (it is an archipelago) nor an "island nation" (it is an unincorporated territory).

the function of a theological bridge concept. By attending to the theological conception of vulnerability, other perspectives may be broadened to consider new ideas about vulnerability and new practices of resilience.

Climate Vulnerability

It is relatively well established that some people are more vulnerable to climate change than others. Discussions of climate change typically include at least some recognition that the harms of climate change will be unequally distributed and, more compellingly, that those most vulnerable to climate harms will disproportionately be those least responsible for contributing to them. This is why climate change is compelling as a matter of justice, beyond any other fundamental environmental–ethical claims. Unequal climate vulnerability is seen as inherently unfair, and this unfairness is compounded by its inverse relationship to culpability. That those with greater culpability and less vulnerability ought to attend to the most vulnerable is seemingly self-evident. This portrayal of the injustice of climate vulnerability is broadly accurate, both descriptively and normatively.

Beyond this association with justice claims, however, the concept of vulnerability has shown itself to be a complex one, providing rich fodder for policy analysis and debate (Brklaich, Chazan, and Bohle 2010; Brooks 2003; Handmer, Dovers, and Downing 1999; Kelly and Adger 2000; Ramsey and Menderson 2011; Usamah et al. 2014). In this regard, I argue that vulnerability—specifically, a theological account of it—may be a useful concept. Historically, the concept has been closely connected with risk and has a long history in a variety of fields related to risk management. With regard to climate change specifically, the use of the concept has primarily been shaped by the IPCC assessment reports and has undergone a number of shifts. Initially, efforts focused on impact assessments, which incorporated two main factors: exposure—that is, the amount of climatic variation to which a system is exposed—and sensitivity—that is, the degree to which the system is affected by that exposure. Impact assessments assume the approach of the natural sciences, applying what has been called the "physical-flows view" or "biophysical vulnerability," concentrating on the flow of matter and energy between system components. In other words, this view focuses on what the climate will do in particular places but not on what people will do in response. In general, most ethical thinking about climate vulnerability seems to share this primary focus on exposure and on the disproportionate exposure of some groups and places to harms from climate change; the effects of climate change are and will be worse in some places than in others.

In contrast, recent consideration of vulnerability incorporates what has been described as an "actor system" view or a focus on "social vulnerability," including among its factors adaptive capacity in addition to exposure and sensitivity and incorporating focused attention on the particular values at risk (Füssel and Klein 2006; Brooks 2003). This latter approach is typified by the IPCC's Third Assessment Report, which specifically names exposure, sensitivity, and adaptive capacity in its definition of vulnerability (Burkett et al. 2014). The Fifth Assessment Report (the most recent one, from 2014) suggests that this increasing complexity of vulnerability has outstripped the usefulness of that earlier definition and accordingly "defines vulnerability simply as the propensity or predisposition to be adversely affected" by climate change. The reference to predisposition should not necessarily be interpreted as a climatic or geographic given but rather as the effect of a combination of causes; thus the report's authors acknowledge that vulnerability is "a multidimensional concept," commending "more attention to the relation with structural conditions of poverty and inequality" (Burkett et al. 2014, 179).

This development of the concept of vulnerability shows a gradual awareness of the complexity of that concept and, additionally, of the importance of resilience. Along with adaptation, resilience is typically posed as the response or solution to vulnerability. Resilience, however, is broader than just adaptation—the Fifth Assessment Report describes resilience as "the ability of a system to respond to disturbances, self-organize, learn, and adapt" (Burkett et al. 2014, 179). In short, resilience is how systems manage vulnerability. In terms of the three factors of vulnerability noted (i.e., exposure, sensitivity, and adaptive capacity), resilience responds to vulnerability primarily through adaptive capacity and sensitivity (sometimes called *internal factors*) rather than exposure (which, as an external factor, is less controllable except through mitigation). So understood, resilience includes not only adaptation (which reacts to disturbances) but also preemptive responses that seek to reduce sensitivity to disturbances before they occur.

Based on this comprehensiveness, I treat resilience as the primary category for describing the goal of reducing vulnerability. Later, the theological account will add to this the notion of resistance to what theologian Elizabeth Gandolfo (2015) calls "the mismanagement of vulnerability" (137).

Vulnerability is not simply the result of innate climatic and geographic differences but also actively produced by structures that make some people and communities more vulnerable; therefore, reducing vulnerability requires not only developing resilience but also actively resisting those structures. At present, however, the focus is on the policy accounts of vulnerability and thus on the goal of resilience.

Resilience is initially an ecological term and can be understood in narrowly biophysical terms, according to the "physical flows" view. But the IPCC notes that resilience is determined by multiple biophysical and social factors, including "poverty, unemployment, quality of housing, or lack of access to potable water, sanitation, health care, and education interacting with land degradation, water stress, or biodiversity loss" (Burkett et al. 2014, 179–80). Consequently, resilience, like vulnerability, is multidimensional, in the terminology of the Fifth Assessment Report, combining human and nonhuman factors and social and biophysical elements. I would also note that, as with vulnerability, the factors of resilience can also be internal or external, whether social or biophysical. An example of an external biophysical factor would be climate variability itself; an internal biophysical factor would be species biodiversity. An external social factor would be the existence of strong centralized forms of support; an internal social factor would be a high degree of internal social capital. Because vulnerability is a result not only of ecological dynamics but also of internal and external social and political dynamics, resilience requires social and political strengths as well as ecological health.

A multidimensional understanding of vulnerability and resilience shifts the focus toward aspects of vulnerability that have previously been overlooked, such as its psychological and spiritual aspects (Clayton, Manning, and Hodge 2014). Unsurprisingly, like its physical impacts, the psychological impacts of climate change are unevenly distributed, and this inequality follows many of the same trajectories, with psychological vulnerability reinforcing physical, political, and economic vulnerability. Communities that are most affected by the physical impacts of climate change—that is, communities that have the most exposure—are naturally more likely to experience adverse psychological effects. Communities with poor infrastructure, high levels of poverty, and inadequate access to health care and other resources will be more susceptible to both physical and psychological impacts (Clayton et al. 2014, 12–13). But psychological impacts also follow their own uniquely unequal logic. Within communities, women, children, and older adults are more susceptible than others to the mental health impacts of climate change (Clayton et al. 2014, 21). Compared with other manifestations of climate change, drought is especially linked to psychological impacts. And communities in which social cohesion, trust, and equality are low are also particularly vulnerable to the psychological effects of climate change. It will be noted that precisely these social factors seem to be eroding in many parts of the world at the same time that the physical impacts of climate change go unheeded by many and US environmental policy turns away from meaningful action.

Just as psychological vulnerability follows its own logic, psychological forms of resilience can function to mitigate some of the most disparate impacts of climate change. A report by the American Psychological Association (APA) and climate change advocacy group ecoAmerica offers practical guidelines for developing psychological and social resilience (Clayton et al. 2014). It notes that both mitigation and adaptation require that people have confidence that climate change can be addressed and responded to effectively, and the authors suggest specific ways of instilling this confidence. More directly related to psychological vulnerability and adaptation, the report's authors recommend steps to strengthen community cohesion and trust and instill a sense of calm and safety, along with particular attention to those populations that are more susceptible to psychological impacts. They note the importance of worldviews and interpretations in making sense of climate change risks and suggest that community actors can be instrumental in creating strong community identity and an overall perspective of optimism. They include faith communities among the actors best suited to take these steps. Similarly, in a study of the impacts of a "super cyclone" on the Brahma Kumaris sect in India, Tamasin Ramsay and Lenore Manderson (2011) argue that religion and spirituality are key components of psychological climate resilience. The specific characteristics they identify of resilience, including meaning making, reflective practice centered on optimistic interpretations of suffering, and "reauthoring" of "self-narratives," echo the findings of the APA/ecoAmerica report (Ramsay and Manderson 2011).

This discussion leads to a final point about this multidimensional understanding of vulnerability and its counterpart, resilience. Assessing vulnerability requires identifying the particular values that are at risk. What, precisely, is vulnerable to climate change, and why should it matter? The policy analyses of vulnerability assess it in terms of a variety of goods and accounts of human flourishing, but as ethicists and philosophers know, flourishing is a category that admits a wide range of definitions. A basic minimum of material well-being is required for flourishing on any account. What I am arguing is that there are societal and cultural goods above and beyond that minimum that are also necessary to a well-lived human life, and these are subject to deliberation and reflection. What aspects of a community and its way of life are we unwilling to risk? Likewise, resilience will have to be defined in terms of some goods and not others. Which values are those on which an assessment of climate resilience depends, and which are we willing to write off as losses? Such is the grim calculus of climate vulnerability; however, these are questions that theological perspectives can help to address.

A Theological Bridge Concept

These latter two points—that resilience can be built through stronger community bonds and that claims about vulnerability involve claims about what communities value—highlight two areas in which religious communities can be directly involved in assessing and managing vulnerability and building resilience. To consider how a theological conception of vulnerability might help communities manage vulnerability, I turn to the concept of "bridge terms" deployed by environmental pragmatists.

Environmental philosopher Bryan Norton (2005) worries that effective environmental ethical discourse is enervated by a lack of "generally recognized vocabulary that makes transparent the connections between ecological outcomes and social values" (38)—that is, by a linguistic inability to connect scientific data with value systems that would make those data meaningful. He suggests that what are needed are bridge terms: "terms that have empirical, operational, and measurable descriptive content and therefore have a connection to the descriptive discourse and the literature of science; but bridge terms also connect to social values and our evaluative discourse by embodying or evoking important social values" (38). Bridge terms bridge the gap between scientific discourse and discussions of value. Norton offers the example of obesity in medicine: it is empirically recognizable but also carries clear evaluative and normative significance. As an environmental pragmatist, Norton is most interested in how such terms function in deliberation about environmental goods; it is left to particular communities to provide the substantive normative content of the

terms. Norton's central bridge term, *sustainability,* is left only loosely—
"schematically"—defined, allowing "real communities" to specify the relevant indicators: "If the people of a community choose indicators associated
with the values they hold dear, and use these indicators to state concrete
sustainability goals with respect to their community, they will in effect be
defining sustainability—for themselves" (40).

Ethicist Willis Jenkins (2013) sees similar discursive potential in sustainability, describing it alternatively as a bridge concept or "middle axiom"
(131). He notes that such ideas "can help open transdisciplinary, cross-cultural dialogue about confronting planetary problems" (140); like Norton, he
is most interested in how "pluralist social practices" enact and drive such
dialogue. Furthermore, Jenkins suggests a distinctive role for theological
engagement with bridge terms like sustainability. Religious and philosophical ethics, he argues, can raise crucial metaethical questions about the
background assumptions that the ideas entail, challenging their presuppo-
sitions and opening the epistemic space to reflect more clearly on the values
at stake.

Implicit in the notion of bridge concepts, including (perhaps especially)
theological ones, is that the communities or disciplines being bridged need
not, and probably would not, have common foundational commitments. In
other words, the scientific or political community need not share the theological commitments of a religious community in order for both communities to benefit from the exchange. If a community articulates (or practices)
its definition of sustainability or, as I will suggest, of vulnerability based on
its foundational commitments, others need not share those commitments
to recognize the validity of that definition and the moral force of its claims.

Appropriately for the Puerto Rican context, this notion of a theological bridge concept shares significant commonalities with an understanding
of Latinx theology as public theology, as described by Benjamín Valentín
(2002). Valentín observes the tendency in Latinx theology toward culturally specific "local" theologies and identity politics, and he urges that this
tendency be balanced by a "public perspective" that engages in "a broader
emancipatory sociopolitical project" (xvi–xxi). He argues for "subaltern
counter*publics,*" the idea of a "more comprehensive arena of discourse and
association in which more limited, yet significant and contextually specific, public spheres infused with diverse values, identities, cultural styles,
and context-specific needs can coexist and, when necessary, unite for the
common good" (124). In other words, Latinx theologies can speak from

particular cultural communities to engage and challenge the broader public discourse. This approach is functionally similar to pragmatist bridge concepts but with a special emphasis on the experiences and identities of marginalized communities. Theological claims drawn from the experiences of these communities can serve to enhance and critique pluralist deliberation, performing the kind of transformative work Jenkins describes.

I suggest that a theological conception of vulnerability can perform precisely this transformative, horizon-broadening function. I will demonstrate that a theological account of vulnerability shares certain resonances with the policy account I have described. These resonances, then, become the basis for vulnerability's usefulness as a theological bridge concept: based on the similarities between the two conceptions of vulnerability, more thorough consideration of the implications of the theological conception might expand our understanding of what climate vulnerability is, does, and can do. Taking theological vulnerability seriously can shift our perception of climate vulnerability such that, instead of victimhood, we see in it agency, power, and claims for justice. On Jenkins's (2013) account, this is what theological bridge concepts do.

To point to resonances or similarities is not to suggest that science endorses the theological conception, or vice versa, or even that the two conceptions are describing precisely the same phenomena—they are not. But there is enough consistent and meaningful overlap between the two to suggest the potential usefulness of vulnerability as a theological bridge concept. Moreover, the practices of vulnerability that I will discuss in the final section of the chapter affirm this usefulness by demonstrating how communities define vulnerability "for themselves" in ways that challenge more straightforward conceptions.

Theological Vulnerability

For Christianity, vulnerability has obvious theological references. Considered theologically, Christian thinking about vulnerability inevitably begins with Christ and the cross—that is, with the vulnerability God assumes in becoming incarnate and suffering. This is not to say that the relationship between Christian communities and vulnerability cannot be considered from other perspectives. It might be evaluated sociologically or ethnographically, for example, examining systematically how beliefs and practices affect vulnerability "from the outside," as it were, as in the aforementioned study of the Brahma Kumaris in India (Ramsay and Manderson 2011). However, for Christian communities themselves, whatever distinctive meaning vulnerability might have is linked to Christ's own vulnerability. If, as I have argued, assessing vulnerability and developing resilience involve deliberative reflection on cultural goods and values, then internal resources like a theological conception of vulnerability are important tools.

We could turn here to any number of theological discussions of the incarnation and kenosis to make this point about theological vulnerability. Feminist theology has been especially fruitful in this area; I will consider the work of two feminist theologians whose sustained and nuanced treatments of vulnerability match the multidimensionality of vulnerability in the IPCC reports discussed. Anglican theologian Sarah Coakley (2002) approaches vulnerability through an astute analysis of the kenotic tradition throughout the history of theology and today. In Christian theology, kenosis refers to the idea (based on Philippians 2:5–11) that in the incarnation, Christ "emptied himself," divesting himself of some or all of his divine nature. In "Kenosis and Subversion: On the Repression of Vulnerability in Christian Feminist Writing," Coakley traces six different meanings of kenosis found in the tradition and notes their incommensurability. Responding to justified feminist concerns about the uses and misuses of kenosis, she argues for a view that sees it neither as a temporary relinquishment of Christ's divine powers nor as a complete identification of divine nature with human limitation (among other options) but rather as the revelation of a God who rejects certain worldly forms of power in themselves and, in so doing, establishes an ethical model of power that is starkly different from common worldly conceptions (10). This interpretation, she suggests, conceives Christ's vulnerability not as a problem or embarrassment for theology to explain away but rather as characteristic of a God who is "humble" and "non-grasping" (10) and thus as a "(special sort of) human strength" (25). Resituating the Christ hymn of Philippians in its original liturgical context, Coakley suggests that this model of strength in vulnerability can be internalized through practices of contemplative prayer (34). The hermeneutical and theological questions here are complex (and ably dealt with by Coakley), but for our purposes, the upshot is a christological view that sees vulnerability not as the negation of power but rather as the true shape of divine and human power. Applied to the present context of climate vulnerability, I suggest that Coakley's christology urges us to look for instances of power precisely in vulnerability itself and inquire how those instances might be developed.

Theologian Elizabeth O'Donnell Gandolfo (2015) describes a similar perspective on vulnerability and considers the practices (including contemplation) whereby it might be enacted. Gandolfo asserts that vulnerability is intrinsic to humans' original condition and, therefore, is the context in which both sin and redemption occur (5). This condition of vulnerability is occasion for anxiety, and so humans seek to manage vulnerability through social structures (136). Societal institutions ("economic, political, cultural, and religious systems of human interaction" [136]) provide a variety of assets that build resilience in the face of vulnerability—physical assets like food and housing, human assets like health and education, social assets like communities and support systems, environmental assets like clean air and water, and existential assets like religion and the arts that help communities understand the world and their place in it.

Given the ubiquity of the "anxiety-induced corruption of egocentrism and parochialism" (Gandolfo 2015, 25), attempts to manage vulnerability inevitably give way to the "communal mismanagement of vulnerability" in social structures of inequality, injustice, and privilege. In turn, these structures exacerbate the problem by producing suffering and increased vulnerability for both the privileged and those "on the underside of privilege" (169).

Gandolfo (2015) suggests that understanding these problems—sin, anxiety, and privilege—through the lens of vulnerability shifts our interpretation of them such that we begin to seek out those existential assets that "offer access to the divine love that consoles and empowers human beings for resilience and resistance in the face of our vulnerable existence" (173)—resilience to bear our original condition of vulnerability, and resistance to the structures of privilege that mismanage that vulnerability.

Drawing particularly on accounts of the paradigmatic vulnerability experienced in motherhood, Gandolfo (2015) identifies three groups of practices that "nurture the growth of divine love in vulnerable human lives and relationships" (266): memory of suffering, contemplative kenosis, and solidarity with vulnerable others. With respect to the first group of practices, she points to the ability of memory to disrupt established conceptions of identity and create new imaginations of an alternative future: "The practice of remembering suffering can empower resilience and resistance insofar as it effects a rejection of received self-understandings that fragment, devalue, and destroy personal and communal identity" (281). In place of these oppressive and alienating identities, memory of past suffering can foster a vision of a new world where suffering is overcome, where "the courage, peace, and compassion of divine love reign" (284). The second family of practices, contemplative kenosis, further nurtures this transforming imagination by allowing persons to experience themselves as loved by God and, more profoundly, as "sacred *loci* of divine presence and activity in the world" (290). Citing Coakley, she describes this contemplation as kenotic because the false identities imposed by others are surrendered as individuals submit their identities and meaning to God—"losing one's life to save it" (292). Finally, both these sets of practices, she says, are fulfilled in solidarity. In contrast to privilege, which is the mismanagement of vulnerability, solidarity is a way of managing vulnerability that reduces the vulnerability of the whole community while requiring some degree of self-sacrifice from some members (300). The new self-identity fostered by memory and contemplation is revealed as a collective identity of "radically social human beings constituted by and called into solidarity with one another" (302). Solidarity empowers persons to recognize their oneness with God, with one another, and with all creation—including, most challengingly, with those who violate vulnerability.

A theological understanding of vulnerability developed along the lines articulated by Gandolfo and Coakley can function as a theological bridge concept in the way that I have described, building on areas of resonance with the science of climate vulnerability to suggest a more expansive conception. To begin with, consider the resonances. On the theological account, there is agency in and around vulnerability, just as there is in the multifaceted policy accounts. Although vulnerability is often understood passively, its social, political, and psychological dynamics introduce a great deal of agency on the part of vulnerable communities. Vulnerable persons and communities exercise agency that defines, shapes, and mitigates their vulnerability. They show varying degrees of social and psychological resilience; they adapt to risks differently. In addition, agency is exercised in producing vulnerability (or, in Gandolfo's [2015] words, mismanaging our innate vulnerability): if communities are more vulnerable based not only on their geographical location but also on their relative social and political locations, then their vulnerability is at least partly a product of someone else's agency. Some groups make others more vulnerable. Unequal climate vulnerability does not simply happen.

A comparison of the two accounts of vulnerability points beyond the simple fact of agency to the power inherent in vulnerability itself. As Gandolfo (2015) emphasizes, and as the other accounts confirm, the seeds of resilience and even of human flourishing are found in vulnerability. In Gandolfo's practices of "the power-in-vulnerability of love," the memory of vulnerability and suffering and the intentional vulnerability of contemplation are themselves resources for creating new identities, new imaginations, and a new solidarity. By way of comparison, the APA/ecoAmerica report speaks of the need for shared meaning, hope, a shared sense of identity, and community trust and solidarity as essential resources for resilience. These are examples of the "existential assets" that Gandolfo describes, and according to the theological account, they are found precisely in vulnerability itself as communities respond to it, reflect on it, pray on it, and act on it.

In this respect, it is worth noting that as feminist theologians, both Coakley and Gandolfo focus on women's experiences of vulnerability and agency—a focus that is particularly appropriate to considerations of climate vulnerability because in most developing countries, women bear both disproportionate vulnerability and disproportionate responsibility for adaptation (Terry 2009; Smith et al. 2014; Olsson et al. 2014). Examining the responses of indigenous communities to climate change, environmental ethicist Elizabeth Allison (2017) articulates a feminist care ethic for climate change that shares many characteristics with the theological account of vulnerability—an emphasis on shared histories and collective identity and action and a value-system that challenges dominant (specifically economic) priorities and attends closely to the most vulnerable. If the theological account urges greater attention to the agency operative within climate vulnerability, it affords particular importance to the experiences of women as both uniquely vulnerable and uniquely powerful.

Finally, a comparison of the two accounts of vulnerability also highlights its duality: climate vulnerability faces in two directions at once, making two kinds of claims. *Vulnerability* refers to both environmental conditions and, at the same time, to social, political, and economic dynamics that exacerbate exposure to climatological risk. Some countries and communities make other countries and communities more vulnerable. Gandolfo (2015) and Coakley (2002) both confirm this duality on theological grounds. Vulnerability may be a part of the human condition, but that innate vulnerability is mismanaged or violated by structures of power and privilege, and vulnerability as a theological claim condemns those structures. In Gandolfo's terms, the response to innate vulnerability is resilience; the response to mismanaged vulnerability is resistance. The roots of both resilience and resistance are in vulnerability itself. When communities claim their vulnerability as the source of their identity and solidarity with one another and God, they thereby reject attempts to mismanage vulnerability through violence and power. Vulnerability can simultaneously build resilience among the most vulnerable and make a justice claim against those who would inflict that vulnerability.

The framework of decolonial thought may be helpful in illuminating this way of looking at vulnerability. Decolonial thought and praxis aim "to delink from the epistemic assumptions common to all areas of knowledge established in the Western world"—specifically, the assumptions of "the colonial matrix of power" (Mignolo and Walsh 2018, 108–9). In other words, challenging colonialism and its neocolonial antecedents requires undermining the epistemologies of colonialism, which underlie all forms of knowledge produced from within that political, economic, and cultural system. This means confronting Enlightenment rationality with alternative ways of knowing and alternative pedagogies—expressing and dramatizing the experience of coloniality, of historical ideological and epistemological marginalization, through stories, rituals, and songs, and elevating indigenous and land-based ways of knowing that emphasize relationality and community (Mignolo and Walsh 2018, 92–95). As Catherine Walsh and Walter Mignolo affirm, "decoloniality was born in the unveiling of coloniality" (6)—the praxis and thought of resistance emerge in the revelation of colonial structures and epistemologies for what they are. Likewise, creative theological expressions of unequal vulnerability make plain the colonialist systems that engender that vulnerability and thus that contain the seeds of resilience and resistance. I will consider some possible examples of such expressions in the next section.

As a theological bridge concept, vulnerability brings normative content and force to the already rich understanding of climate vulnerability offered by the policy accounts. The theological conception draws attention to the power resident in vulnerability itself and articulates practices that develop resilience from within vulnerability. At the same time, it makes clear that vulnerability is bilateral, that the same memories and identities that nurture solidarity and resilience among the vulnerable also indict the structures of privilege that violate human vulnerability and those who benefit from them, calling for resistance. Recalling Norton's (2005) claim that communities define bridge concepts for themselves, this theological account begins to provide definition to the idea of vulnerability as a bridge concept. From the environmental pragmatists' perspective, it remains to consider how real communities make such a conception of vulnerability operational in concrete practices.

Practicing Island Vulnerability

This chapter opened with the example of Puerto Rico in Hurricane Maria as particularly illustrative of the multidimensionality of climate vulnerability. I now return to that example, along with illustrations from other islands (the Philippines and the Maldives) to consider practices that I believe illustrate the account of vulnerability I have been developing. First, however, I address the special climate vulnerability of islands and island nations.

As Hurricane Maria and its aftermath amply demonstrated, islands like those that comprise Puerto Rico are uniquely vulnerable to the effects of climate change. The IPCC reports that although islands are diverse and their vulnerability is therefore varied, they are highly vulnerable to climate change-related stressors (Nurse et al. 2014). These stressors can be considered according to the three components of climate vulnerability described earlier. Consequently, in terms of exposure, islands are more likely to experience many of the effects of climate change. These include sea-level rise and its attendant problems of flooding, shoreline change and territory loss, and salinization of fresh groundwater resources but also coral reef bleaching from rising ocean temperatures and increased occurrences of extreme weather events. Sensitivity is also heightened and adaptive capacity reduced in island contexts. Species extinction and occurrences of invasive species are likely to be higher owing to islands' limited area and isolation; these same factors also contribute to greater human health risks, and economic impacts are more pronounced because of island economies' dependence on ocean ecosystems (Nurse et al. 2014; Kelman and West 2009). Adaptive capacity is further undermined by the economic conditions of many small islands. Small size and isolation lead to high per capita infrastructure costs (Nurse et al 2014, 1625). Most island economies are heavily dependent on strategic imports and thus are more volatile than economies without that dependence. Governance issues on many Pacific island nations may also affect adaptive capacity (Nurse et al. 2014, 1625).

Puerto Rico faces its own unique political and economic challenges. As a territory of the United States, it is subject to federal taxation and contributes to the Federal Emergency Management Agency. It is also subject to US trade policies like the Jones Act of 1920, which limits the goods (including disaster relief) available to it (Chokshi 2017). At the same time, many believe that it received less federal support than other parts of the United States after the devastating hurricanes of 2017 (Rodríguez-Díaz 2018; Robles, Alvarez, and Fandos 2017). Even before Hurricane Maria, it was in the grip of a 10-year recession exacerbated by the peculiarities of its unique political status. As of this writing, Puerto Rico was $70 billion in debt, with 45 percent poverty and 12 percent unemployment. Before Maria, its physical infrastructure was already in a precarious state. Mental illness was already increasing before the hurricane, and trauma combined with lack of access to medicine and care has led to a burgeoning mental health crisis after it (Dickerson 2017). These social and economic factors have combined with population growth and environmental degradation to create a situation of chronic vulnerability (Maldonado 2009). As a densely populated island territory, Puerto Rico faces an additional challenge: because internal mobility is severely limited, other adaptive strategies are required (Maldonado 2009, 201). Finally, the aforementioned gender disparities with respect to vulnerability and adaptation are borne out in the case of Puerto Rico—women are especially vulnerable, and their agency is especially important for building resilience (Maldonado 2009, 210).

These particular vulnerabilities notwithstanding, some have suggested that islands and island nations also have distinctive characteristics that may enhance resilience. These advantages include "tight kinship networks, unique heritage, a strong sense of identity and community, creativity for sustainable livelihoods, remittances from islander diasporas supporting life on [small island developing states], and local knowledge and experience of dealing with environmental and social changes throughout history" (Kelman and West 2009, 2). Without denying the aforementioned variability among island nations, research suggests that local knowledge and traditional oral narratives may be essential to developing and implementing adaptation strategies (Kelman and West 2009; Janif et al. 2016).

In this context of distinctive vulnerability, a theological conception of vulnerability highlights some practices that mobilize the power of vulnerability. In this section, I consider primarily the response of the Episcopal Church of Puerto Rico (Iglesia Episcopal Puertorriqueña) (*a*) because that is the tradition with which I am most familiar, (*b*) because that church's response is readily available in a number of online media sources, and (*c*) because its response enacts many of the ideas described by Gandolfo (2015). Nonetheless, the examples of other religious communities seem to draw on similar themes to make sense of the devastation and suffering (Shellnut 2017). I also consider recent (secular) calls for statehood and the (secular) examples of the Philippines and the famous underwater cabinet meeting in the Maldives. As noted, the usefulness of a theological bridge concept is not necessarily limited to religious communities.

Again, the perspective of decolonial thought is helpful in examining these examples. The examples each offer some form of the pluralist, counterhegemonic praxis and discourse that approach affirms. The responses show creativity and versatility as they challenge dominant narratives and amplify marginalized voices. They communicate the power of vulnerability through images, liturgy, narrative, and caring action in community.

The response of the Episcopal Church of Puerto Rico to Hurricane Maria and its aftermath does not explicitly invoke vulnerability, nor does the church speak of kenosis, contemplation, or solidarity. Nonetheless, I contend that in its overall response, the church exemplifies Gandolfo's (2015) practices of vulnerability. Since the hurricane, the Episcopal Church, which already had an established health-care service, has created mobile "pop-up" clinics to provide health care, along with food, water, and hygiene items, for the territory's devastated and isolated communities. In conjunction with these services, the bishop of the diocese, Rafael Morales, has traveled to remote villages to provide religious services and fellowship. Priests in the diocese have also overcome impassable roads to visit villagers' homes to provide pastoral care, even when priests' homes (and the diocesan office in San Juan) were without electricity for months (Disaster News Network 2018). The most vulnerable in the community are being sought out and cared for, as the research on social and psychological vulnerability urges.

Most telling is the theology that the church invokes to frame its response. The diocese describes itself as "the diocese of hope," and it has used the disaster as an opportunity to reconceive itself as a missionary diocese. This missionary character is expressed in a sustained focus on face-to-face ministry in local communities, in conjunction with the health clinics. In his regular podcast, the bishop characterizes this ministry as an openness to encountering God in others, in communities, and in daily life, as well as an openness to the mystery of God (Iglesia Episcopal Puertorriqueño 2017). These attempts to make meaning of suffering—by reimagining the situation as one of hope, by emphasizing and responding to both God's inscrutability and God's presence in one another, by nurturing shared identity and solidarity in the face of hardship—all reflect the main themes of

Gandolfo's theology of power in vulnerability. They also point to the ability of theological discourse to answer the fundamental questions raised earlier: What values and qualities are most essential to community identity and, therefore, most worthy of preservation?

Some of the most poignant manifestations of this theological vulnerability have occurred around the celebration of Ash Wednesday and the beginning of the season of Lent. Ash Wednesday uses ashes placed on the forehead as a reminder to Christians of their mortality and dependence on God. In this vein, Bishop Morales describes the Church's practice of "ashes to go," the imposition of ashes outside, on passers-by and even motorists, as a reminder of God's desire to be present in daily life. Similarly, in his first weekly podcast during Lent, the bishop invites his listeners to use the season as a time to focus on openness to God's call for their lives (Iglesia Episcopal Puertorriqueño 2017). And one priest's Lenten meditation explicitly connects the ashes of Ash Wednesday with the traditional woodstoves that have made a resurgence in the post-Maria, de-electrified landscape, and then to vulnerability: "Ashes are a sign of the unbreakable commitment to our own personal growth. But this growth must come not from a position of power and pride, but rather from our own poverty and fragility . . . where we are naked before ourselves and before God; free from all masks, from all the expectations and all the roles that we assume before society" (Morales 2018). This reflection speaks profoundly of the relation of climate vulnerability to the power of theological vulnerability. A final Ash Wednesday vignette indicates the power of the human condition of vulnerability to foster solidarity. 152 days after power had been cut off to the diocesan office, contractors from the mainland restored the electrical connection. They waited patiently through the conclusion of the Ash Wednesday observance, and then asked the bishop to impose ashes (Iglesia Episcopal Diócesis de Puerto Rico 2018). This symbol of shared mortality gave expression to intercultural solidarity in the face of the devastation of climate change.

In the context of the dual nature of vulnerability that I have noted, these actions demonstrate vulnerability as resilience: within vulnerability itself are resources for strengthening communities and cultivating the assets, especially the existential assets, necessary for greater resilience. At the same time, Puerto Rico illustrates vulnerability as resistance—unequal vulnerability stands as an indictment of the structures that mismanage vulnerability and seek the security of some at the expense of others. The increased strength of the movement for statehood for Puerto Rico reflects this duality. In the face of the aforementioned structural inequalities that characterize Puerto Rico's territorial status—that is, the political burdens it bears without receiving commensurate benefits—the already increasingly forceful call for statehood was given new strength (Hulse 2018).[2] When the 2017 tax reform legislation was projected to eliminate incentives for business investment in the territory, this drew further attention to Puerto Rico's uniquely precarious position. In this example, it is not simply the case that heightened climate vulnerability is the occasion for the charge of injustice and demands for structural change but rather that such vulnerability has become the reason for such claims.

A study of the resilience of two villages in the Philippines confirms these anecdotal observations from Puerto Rico (Usamah et al. 2014). Researchers found that residents of the villages reported a high perception of social resilience. Key characteristics of this resilience were identified as trust among community members; social cohesion, which includes common values, solidarity, and identity; high levels of community involvement; and effective formal and informal modes of communication. They note that the church is an important site for the strengthening of social bonds (Usamah et al. 2014, 184). Many of the same characteristics of resilience in vulnerability identified by both the policy and theological accounts are borne out in the villagers' experience.

A final example of vulnerability as resistance, while dramatic, is also ambiguous. The famous underwater cabinet meeting of the Republic of Maldives is a striking example of the convicting power of vulnerability. When then-president Mohamed Nasheed and 11 cabinet members donned scuba gear and carried waterproof documents and markers to hold a cabinet meeting underwater in 2009, they drew the world's attention to the existential threat posed by present and future sea-level rise to islands and island nations. They mobilized the island nation's unique vulnerability as an accusation against the fossil fuel–burning developed world (Henson 2014, 151). Nasheed, the country's first democratically elected president, was eventually ousted in an alleged coup amid political unrest and charges of being anti-Islamic based on his pro-Western policies (Aljazeera News 2012). It is not clear that there is any direct connection between Nasheed's stance

---

2. It is worth noting that in the time between writing and publication, this energy seems to have diminished significantly. Like the Maldives example, this change illustrates what might be called the vulnerability of vulnerability.

on climate change and his ouster; however, in the account of climate vulnerability I have described, political instability and climate vulnerability reinforce one another. In the end, vulnerability is risky.

After all, to be vulnerable is to be weak—susceptible. Whatever power vulnerability may hold, to become vulnerable is to expose oneself to risk. Stories like the underwater cabinet meeting suggest the danger inherent in vulnerability. Emphasizing vulnerability, as President Nasheed did, may only exacerbate the risk. One study of Puerto Rican communities following hurricanes Hugo in 1989, Hortensia in 1996, and Georges in 1998 concluded that, for all that one may hope to find the seeds of resilience, "the shortages that these communities suffered during and after the hurricanes were more difficult to manage than had been expected. The data show that the scarcity of the resources and income necessary for basic survival together with the gradual deterioration of the physical environment limited the capacity to absorb the changes caused by recent hurricanes" (Maldonado 2016). Some examples of climate vulnerability may point to hidden strength; others may simply point to weakness.

The theological perspective confirms the risks inherent in vulnerability. The self-emptying of the incarnation leads to the cross. This is precisely the basis of many feminists' worries about kenotic theology—to enjoin vulnerability on the already vulnerable is simply to invite further suffering. Nevertheless, as the scholars cited in this chapter attest, the turn to vulnerability does not support the production of unequal vulnerability but rather directly challenges those structures and the conceptions of power that support them.

Worries about the vulnerability of vulnerability do not finally disprove its potential for resistance and resilience. The theological conception of vulnerability suggests practices and assets for resilience within vulnerability that scientific and policy-oriented accounts may overlook. As Gandolfo (2015) argues, vulnerability is part of the human condition. It is not overcome in this life, and a life without the risks of vulnerability would, she contends, be missing essential factors of human flourishing. The theological account proceeds not by negating the risks inherent in vulnerability but rather by showing how those risks might best be managed for the benefit of all by using the characteristics of vulnerability itself. Examples like Puerto Rico illustrate that vulnerability, viewed from a theological perspective, does indeed offer resources for building resilience that correspond to the factors of resilience identified in scientific and policy-oriented literature.

That these resources do not always, or even typically, overcome the danger and weakness inherent in vulnerability does not negate their potential to do so in some cases or, accordingly, the need to cultivate them.

Conclusion

I have argued that approaching vulnerability as a theological bridge concept reveals certain assets that can be mobilized and developed to build resilience and resistance. Vulnerability can serve as a bridge concept in this way because of the significant overlap that exists between the scientific and policy-oriented account of climate vulnerability I have described and the theological conception of vulnerability indicated by theologians like Coakley (2002) and Gandolfo (2015). The assets identified—shared identity and memory, trust and solidarity, and indictment of the mismanagement or violation of vulnerability in structures of injustice and privilege—and the practices Gandolfo describes to cultivate these assets bear close resemblance to the components of resilience in the scientific and policy-oriented literature. At the same time, the theological conception of vulnerability expands and enhances these components, adding a call to resistance and giving the concept depth and concreteness in community practices like those of the church in Puerto Rico.

It bears repeating that this argument is in no way intended to suggest that vulnerability itself is good, especially as it is implicated in structures of injustice, or to enjoin vulnerability on already vulnerable communities, nor do I mean to imply that those communities that are unable to develop resilience are somehow to blame for this. I concur with Gandolfo that some forms of vulnerability seem to be an inescapable fact of human life, necessary for human flourishing. But much of what we identify as vulnerability—including, to a large degree, climate vulnerability—is implicated in unjust structures and, therefore, to be resisted. My goal in this chapter is to examine and commend strategies for reducing vulnerability, insofar as it can be reduced, and resisting its violation, insofar as that can be resisted. Specifically, my hope is that those interested in climate vulnerability might learn from the theological conception and, based on this learning, might take note of and seek to cultivate practices that build resilience from within vulnerability. In particular, if resilience and resistance emerge from within experiences of vulnerability, attention to the most vulnerable communities is crucial to developing response strategies to climate change.

Climate change challenges existing modes of ethical reasoning and valuing; it calls for moral creativity and boundary-crossing explorations (Jenkins 2013, 95). Theological bridge concepts like vulnerability can drive such exploration, drawing on the rich history of theological thought about the human condition while speaking meaningfully to human life in the age of climate change. Attending to these concepts can spark creativity by revealing the unexpected ways communities define and engage their values in the concrete struggle to survive and flourish in the face of environmental threats.

References

Aljazeera News. 2012. "Maldives President Quits after Protests." Accessed February 23, 2018. https://www.aljazeera.com/news/2012/2/7/maldives-president-quits-after-protests.

Allison, Elizabeth. 2017. "Toward a Feminist Care Ethic for Climate Change." *Journal of Feminist Studies in Religion* 33 (2): 152–58.

Brklacich, Mike, May Chazan, and Hans-Georg Bohle. 2010. "Human Security, Vulnerability, and Global Environmental Change." In *Global Environmental Change and Human Security*, edited by Richard Anthony Matthew, 35–51. Cambridge, MA: MIT Press.

Brooks, Nick. 2003. "Vulnerability, Risk and Adaptation: A Conceptual Framework." Norwich, UK: Tyndall Centre for Climate Change Research.

Burkett, Virginia R., Avelino G. Suarez, Marco Bindi, Cecilia Conde, Rupa Mukerji, Michael J. Prather, Asuncion L. St. Clair, and Gary W. Yohe. 2014. "Point of Departure." In *Climate Change 2014: Impacts, Adaptation and Vulnerability. Part A: Global and Sectoral Aspects. Contribution of Working Group II to the Fifth Assessment Report of the Intergovernmental Panel on Climate Change*, edited by Christopher B. Field et al., 169–94. Cambridge, UK: Cambridge University Press.

Chokshi, Niraj. 2017. "Would Repealing the Jones Act Help Puerto Rico?" *New York Times*, October 24, 2017. https://www.nytimes.com/2017/10/24/us/jones-act-puerto-rico.html.

Clayton, S., C. M. Manning, and C. Hodge. 2014. "Beyond Storms and Droughts: The Psychological Impacts of Climate Change." Washington, DC: American Psychological Association and ecoAmerica. https://ecoamerica.org/wp-content/uploads/2014/06/eA_Beyond_Storms_and_Droughts_Psych_Impacts_of_Climate_Change.pdf.

Coakley, Sarah. 2002. *Powers and Submissions: Spirituality, Philosophy and Gender*. Oxford, UK: Wiley-Blackwell.

Dickerson, Caitlin. 2017. "After Hurricane, Signs of a Mental Health Crisis Haunt Puerto Rico." *New York Times*, November 13, 2017. https://www.nytimes.com/2017/11/13/us/puerto-rico-hurricane-maria-mental-health.html.

Disaster News Network. 2018. "Puerto Ricans 'Have a Good Heart.'" Accessed February 22, 2018. http://www.disasternews.net/news/article.php?articleid=5833.

Füssel, Hans-Martin, and Richard J. T. Klein. 2006. "Climate Change Vulnerability Assessments: An Evolution of Conceptual Thinking." *Climatic Change* (3): 301–29.

Gandolfo, Elizabeth O'Donnell. 2015. *The Power and Vulnerability of Love: A Theological Anthropology*. Minneapolis, MN: Fortress.

Handmer, J. W., S. Dovers, and T. E. Downing. 1999. "Societal Vulnerability to Climate Change and Variability." *Migration and Adaptation Strategies for Global Change* 4:267–81.

Henson, Robert. 2014. *The Thinking Person's Guide to Climate Change*. Rev. ed. Washington, DC: American Meteorological Society.

Hernández, Arelis R. 2018. "GWU Experts Tapped to Review Hurricane Maria Death Toll in Puerto Rico." *Washington Post*, February 22, 2018. https://www.washingtonpost.com /national/gwu-experts-tapped-to-review-hurricane-maria-death-toll-in-puerto -rico/2018/02/22/7f0a1c94-17e7-11e8-8b08-027a6ccb38eb_story.html.

Hulse, Carl. 2018. "Advocates of Puerto Rico Statehood Plan to Demand Representation." *New York Times*, January 9, 2018. https://www.nytimes.com/2018/01/09/us/politics /advocates-of-puerto-rico-statehood-plan-to-demand-representation.html.

Iglesia Episcopal Diócesis de Puerto Rico. 2018. "Después de 152 días sin servicio de energía eléctrica en el Centro Diocesano de la Iglesia Episcopal Puertorriqueña, contratistas de la Autoridad de Energía Eléctrica de Puerto Rico (AEE) provenientes de los Estados Unidos restablecieron el servicio. Los trabajadores pacientemente esperaron la culminación de nuestro acto litúrgico para solicitar del Señor Obispo Rvdmo. Rafael Morales oración y la imposición de las cenizas. ¡SEÑOR, TÚ NOS LLAMAS!" Facebook, February 14, 2018. https://www.facebook.com/episcopalpr/posts/1599837940051770.

Iglesia Episcopal Puertorriqueño. 2017. *Senderos*, December 24, 2017. Online audio recording.

Janif, Shaiza Z., Patrick D. Nunn, Paul Geraghty, William Aalbersberg, Frank R. Thomas, and Mereoni Camailakeba. 2016. "Value of Traditional Oral Narratives in Building Climate-Change Resilience: Insights from Rural Communities in Fiji." *Ecology and Society* 21 (2). http://dx.doi.org/10.5751/ES-08100-210207.

Jenkins, Willis. 2013. *The Future of Ethics: Sustainability, Social Justice, and Religious Creativity*. Washington, DC: Georgetown University Press.

Kelly, P. M., and W. N. Adger. 2000. "Theory and Practice Assessing Vulnerability to Climate Change and Facilitating Adaptation." *Climatic Change* 47 (4): 325–52.

Kelman, Ilan, and Jennifer J. West. 2009. "Climate Change and Small Island Developing States: A Critical Review." *Ecological and Environmental Anthropology* 5 (1): 1–16.

Maldonado, Zaibette. 2009. "Vulnerabilidad y Resiliencia Ante Las Amenazas En El Municipio de Loíza, Puerto Rico." *Revista Umbral* 1:199–217 (translation mine).

Mignolo, Walter D., and Catherine E. Walsh. 2018. *On Decoloniality: Concepts, Analytics, Praxis*. Durham, NC: Duke University Press.

Morales, D. Francisco. 2018. "Miércoles de Ceniza." In *Reflexiones de Cuaresma*. Oficina Canónigo de Evangelismo Pastoral y Misión, Iglesia Episcopal Puertorriqueña (online booklet). https://goo.gl/11tMS1.

Norton, Bryan G. 2005. *Sustainability: A Philosophy of Adaptive Ecosystem Management*. Chicago: University of Chicago Press.

Nurse, Leonard A., Roger F. McLean, John Agard, Lino Briguglio, Virginie Duvat-Magnan, Netatua Pelesikoti, Emma Tompkins, and Arthur Webb. 2014. "Small Islands." In *Climate Change 2014: Impacts, Adaptation, and Vulnerability. Part A: Global and Sectoral Aspects. Contribution of Working Group II to the Fifth Assessment Report of the Intergovernmental Panel on Climate Change*, edited by Christopher B. Field et al., 1613–54. Cambridge, UK: Cambridge University Press.

Olsson, L., A. Tschakert, A. Agrawal, S. H. Eriksen, S. Ma, L. N. Perch, and S. A. Zakieldeen. 2014. "Livelihoods and Poverty." In *Climate Change 2014: Impacts, Adaptation, and*

Vulnerability. Part A: Global and Sectoral Aspects. Contribution of Working Group II to the Fifth Assessment Report of the Intergovernmental Panel on Climate Change, edited by Christopher B. Field et al., 793–832. Cambridge, UK: Cambridge University Press.

Ramsay, Tamasin, and Lenore Manderson. 2011. "Resilience, Spirituality and Posttraumatic Growth: Reshaping the Effects of Climate Change." In *Climate Change and Human Well-Being*, edited by Inka Weissbecker, 165–184. New York: Springer.

Robles, Frances. 2017. "Official Toll in Puerto Rico: 64. Actual Deaths May Be 1,052." *New York Times*, December 8, 2017. https://www.nytimes.com/interactive/2017/12/08/us /puerto-rico-hurricane-maria-death-toll.html.

Robles, Frances, Lizette Alvarez, and Nicholas Fandos. 2017. "In Battered Puerto Rico, Governor Warns of a Humanitarian Crisis." *New York Times*, September 25, 2017. https://www.nytimes.com/2017/09/25/us/puerto-rico-maria-fema-disaster-.html.

Rodríguez-Díaz, Carlos E. 2018. "Maria in Puerto Rico: Natural Disaster in a Colonial Archipelago." *American Journal of Public Health* 108 (1) (January 2018): 30–31.

Schwartz, Emma. 2018 "Hurricane Maria's Uncertain Death Toll in Puerto Rico." Frontline, August 17, 2018. https://www.pbs.org/wgbh/frontline/article/hurricane-marias -uncertain-death-toll-in-puerto-rico/.

Shellnutt, Kate. 2017. "How Do You Solve a Theological Problem Like Maria?" *Christianity Today*, September 20, 2017. http://www.christianitytoday.com/news/2017/september /how-solve-problem-maria-irma-hurricane-theodicy-caribbean.html.

Smith, Kirk R., Alistair Woodward, Diarmid Campbell-Lendrum, Dave D. Chadee, Yashudi Honda, Qiyong Liu, Jane M. Olwoch, Boris Revich, and Rainer Sauerborn. 2014. "Human Health: Impacts, Adaptation, and Co-benefits." In *Climate Change 2014: Impacts, Adaptation, and Vulnerability. Part A: Global and Sectoral Aspects. Contribution of Working Group II to the Fifth Assessment Report of the Intergovernmental Panel on Climate Change*, edited Christopher B. Field et al., 709–54. Cambridge, UK: Cambridge University Press.

Terry, Geraldine. 2009. "No Climate Justice without Gender Justice." *Gender and Development* 17 (1): 5–18.

Usamah, Muhibuddin, John Handmer, David Mitchell, and Iftekhar Ahmed. 2014. "Can the Vulnerable Be Resilient? Co-existence of Vulnerability and Disaster Resilience: Informal Settlements in the Philippines." *International Journal of Disaster Risk Reduction* 10:178–89. https://doi.org/10.1016/j.ijdrr.2014.08.007.

Valentin, Benjamin. 2002. *Mapping Public Theology: Beyond Culture, Identity, and Difference.* Harrisburg, PA: Bloomsbury T&T Clark.

ANDREW R. H. THOMPSON is Assistant Professor of Theological Ethics at the School of Theology at the University of the South (Sewanee), where he also directs the Alternative Clergy Training at Sewanee program. His research and teaching focus on environmental ethics and ecotheology. His book, *Sacred Mountains: A Christian Ethical Response to Mountaintop Removal*, was published by University Press of Kentucky in 2015.

9

RESILIENCE AND RELIGION

What Does Civic Diplomacy Have to Do with It?

Roger-Mark De Souza

JOHN FEFFER AND NANCY SNOW (2016) INDICATE THAT initial US government public diplomacy efforts to share the universality of US values with the world were focused on promoting the national security interests of the United States. They suggest that civic diplomacy would be a more promising approach if focused on "second-track" people-to-people exchanges that promote mutual understanding across countries. Since the 1950s, civic diplomacy along these lines has historically been facilitated by the involvement of citizens as "diplomats" serving the mutual interests of respective communities. Internationally, this civic diplomacy has been driven by the United States through a national initiative embodied by Sister Cities International (SCI 2021). SCI was conceived at President Eisenhower's 1956 People-to-People Conference as a network of citizen diplomats who would champion peace by fostering bonds between different communities around the world. President Eisenhower reasoned that if people from diverse cultures could understand, appreciate, and celebrate their differences while building partnerships, their efforts would lessen the chance of new conflicts.

Civic Diplomacy and Global Engagement

Dyadic city-to-city relationships such as Sister City partnerships typically pair communities of different countries and facilitate bilateral flows of information, people, commerce, and other resources. As such, Sister City relationships historically serve as a goodwill platform for countries to undertake activities for the enhancement of peaceful and cooperative relationships (Chung and Mascitelli 2009, 228). These relationships are considered civic diplomacy insofar as they involve cooperation among citizens at a city or community level, representing an established diplomatic relationship with other entities far outside their jurisdictions. Cities have developed this form of international involvement that is sustained and effective over time (Leffel and Acuto 2017, 13; Chung and Mascitelli 2009, 228).

A Sister City (or, for that matter, "sister county" or "sister state") relationship is a broad-based, long-term partnership between two communities in two countries. The partnership is officially recognized after the highest elected or appointed officials from both communities sign off on an agreement, typically a memorandum of understanding. A Sister City organization may have any number of sister cities, with community involvement ranging from a half dozen to hundreds of volunteers. In addition to volunteers, Sister City organizations may include representatives from nonprofits, municipal governments, the private sector, and other civic organizations. Sister City relationships offer the flexibility to form connections between communities that are mutually beneficial and that address issues that are most relevant for partners.

On the most basic level, Sister City relationships allow citizens to exchange ideas, gain international perspectives, and develop a more nuanced understanding of global issues. However, many Sister City programs go much further, lending economic growth and development to both their home and partner communities and humanitarian support and stability to volatile regions. Through the lens of civic diplomacy, a Sister City program enables citizens—and, by extension, their community—to become directly involved in international relations in a distinct and meaningful way, bringing long-term benefits to both the US community and its partners abroad.

Civic Diplomacy and Religious Social Capital

Civic diplomacy relies on the ability to generate social capital—that is, resources that accrue from social networks. Typically, social capital is generated through intermediary organizations such as clubs, organizations, associations, and churches. Religious social capital may be conceived as the social resources available to individuals and groups through their social connections with a religious community. Key components include group membership, social integration, values and norms, bonding and bridging trust, and social support (Maselko, Hughes, and Cheney 2011). Some studies have noted that individuals who are active in religious congregations tend to be more civically engaged than either religiously unaffiliated adults or inactive members of religious groups (Pew Research Center 2019). Researchers such as Aubrey Cox, Melissa Nozell, and Imrana Alhaji Buba (2017) have even examined how religious social capital may be leveraged for other civic engagement opportunities including peace building and youth mobilization.

At a local level, Spickard (2020) analyzed two case studies to demonstrate how church-based activism could generate public participation. He posited that a series of values-based organizing tactics strengthened civic engagement. These tactics and strategies included leveraging theological reflection to build a community of actors, increasing the base of actors across religious denominations and racial identity lines, and cementing new relationships by forging collaboration with environmentalists and authorities.

The Roots of Sister City Civic Diplomacy

Since its inception, SCI has played a key role in renewing and strengthening important global relationships between countries and communities that may or may not have had contentious histories. In the aftermath of World War II, relationships between Japan and the United States were extremely tense. There was animosity on both sides and a reluctance to form any sort of connection on a large scale. However, individual cities and towns in the United States and Japan were able to move past the postwar political climate and forge meaningful relationships with one another built on goodwill and common interests.

The first Sister City connection between the United States and Japan was initiated in 1955 between St. Paul, Minnesota, and Nagasaki, Japan, on the anniversary of the Pearl Harbor attack. In visiting Peace Park in Nagasaki in 1990, St. Paul mayor Jim Scheibel noted, "It is important that we honor the memory of war victims, in our country as well as Japan. The best way to do that is to work to make sure that this kind of war and destruction never happens again. That's what Sister City relationships and people-to-people diplomacy are all about" (SCI 2006).

With a unique civic diplomacy approach to global development, the formation of long-term Sister City relationships quickly gained recognition and momentum in the United States. By the mid-1960s, city affiliations realized that their diverse efforts needed proper coordination through a civic diplomacy intermediary organization. In 1967, SCI—then known as the Town Affiliation Association of the United States—was created. Its newly elected president, Ambassador George V. Allen, recognized that there was a need for a more formal process for the establishment of a Sister City relationship.[1] The Town Affiliation Association created a membership organization that would facilitate communication, teach best practices, and, in time, become a repository of knowledge available for any city interested in the benefits of civic diplomacy.

Civic Engagement and Cultural Ambassadors

This approach to diplomacy was premised on the belief that leveraging a civic approach to diplomatic relations would introduce citizens to other countries and cultures through face-to-face interactions and cultural exchanges. Such efforts could then, in turn, lead to a better understanding of those people and, ultimately, a better and more peaceful world. From a public diplomacy perspective, SCI and its associated network of Sister City organizations act as cultural ambassadors for the United States and foreign entities, bringing communities together and introducing them to new ideas and cultures.

One notable example of this approach is embodied in the strong partnership that SCI forged with China in the 1970s. When the current president of China, Xi Jinping, first came to the United States as a young man, it was through a Sister States program. Rather than basing his impressions of the United States solely on the US government's policy, he was able to connect with the United States on a more human level through his experience with SCI. President Xi Jinping returned to Muscatine, Iowa, in 2012, twenty-seven years after his first visit when he was a young provincial official from Hebei province.[2] The president said of his experience in Iowa, "You can't even imagine what a deep impression I had from my visit 27 years ago to Muscatine, because you were the first group of Americans that I came into contact with. My impression of the country came from you. For me, you are America" (Grote 2012).

---

1. A more in-depth look at the first 50 years of SCI (2006) history can be found in the book *Peace through People: 50 Years of Global Citizenship*.

2. At the time of his return visit, Xi Jinping was vice president of China (see Johnson 2012).

In the mid-1970s, there was general interest in expanding programing. The Town Affiliation Association began the School Affiliation Program. Through this program, youth gained greater sensitivity toward other cultures and a broader global perspective. In the Sister City collaboration between Oakland, California, and Fukuoka, Japan, the two cities spent a school year exchanging artwork and conducting workshops on Japanese culture. Additional programs at that time focused on basic urban problems such as water and sanitation, health, housing, education, and transportation. The Technical Assistance Program (TAP) began in 1977 and worked to create training programs between sister cities to increase employment; to establish cooperatives, credit unions, and other small-scale organizations; and to provide the opportunity for more positive interactions between sister cities. Development agencies throughout the world realized that industrializing countries experienced many of the same urban problems as developed nations. The Sister City movement provided a mechanism for communities to share their experiences and best practices, creating stronger and more beneficial relationships. Communities quickly learned that TAP provided opportunities that would not have otherwise been available. A city project to improve surface drainage, for example, would aid the urban poor. The citizens would gain better sanitation and possible employment from the project.

Municipal and Federal Partnerships

During the 1980s, SCI developed a focus on municipal twinning. Mayors began to focus on forming relationships that offered technical assistance in municipal development. Similar to TAP, these exchanges worked on citywide issues such as solid waste management, urban planning, emergency response training, and emergency management. Cities also began to utilize their Sister City relationships to improve their international trade development and open themselves up to the outside world. In one joint venture, for example, a Baltimore, Maryland, business sent engines to a business in Xiamen, China. The business in China used the engines in excavating equipment and forklift manufacturing (SCI 2006).

Today, civic engagement as a component of official diplomacy is supported and encouraged through the US Department of State's (2021a) Bureau of Educational and Cultural Affairs, where such civic participation is considered an inherent part of foreign policy. Civic diplomacy is believed to help share US ideas and beliefs in democracy, civic society, human rights, and freedom of speech and religion. These State Department programs build social capital and cultivate citizen ties across geographies that lead to long-term networks and personal relationships that are central to promoting US national security interests and values. Other State Department initiatives have helped build social capital. An evaluation of six State Department International Religious Freedom Act of 1998 projects examined ways that such projects helped leverage approaches, including civic diplomacy, to assist in promoting religious freedom. Despite not being able to fully institutionalize all interventions or achieve sustainable impact for religious freedoms in conflict-affected areas of the world, these projects were found to build social capital and informal networks, increasing the potential for broader social change and building trust to rehumanize the "other" (US Department of State 2021b).

From Civic Diplomacy to Broader Impacts

In May 1989, the first Sister Cities conference between the United States and the Union of Soviet Socialist Republics (USSR) was held in Tashkent, USSR (now in Uzbekistan). Sister Cities in the United States deployed almost five hundred delegates from more than forty cities across the country to attend this conference. Similarly, an unprecedented 220 Soviet officials and citizens traveled to the United States in September 1991 to attend another US–USSR Sister Cities conference in Cincinnati, Ohio—the largest ever gathering of Soviet citizens in the United States. With the USSR's troubled political and economic situation, the delegates discussed important topics, including communications development, citizen involvement, education, and further opportunities for affiliation. The conferences sought to encourage international understanding and communication, as well as mutual respect, among ordinary citizens and to explore future business and trade partnerships.

Although these events preceded the collapse of the USSR and independence of the former Soviet republics at the end of 1991, they provided ample opportunity for the formation of friendships and partnerships between US and Soviet business communities. Many Sister City members took advantage of the growing world economy and developed lucrative business agreements with their partners. Vermont's Ben and Jerry's Ice Cream Company, for example, started a factory in Karelia, Russia. The company offered the same profit-sharing opportunities to its Russian employees as their American counterparts. Although business relationships were not the primary goal of SCI, they became a natural by-product of Sister City exchanges.

SCI has continually improved diplomatic relationships at watershed moments. SCI's Muslim World Partnership Initiative created more than one hundred relationships between American and Islamic communities. Through the "Partners for Peace" initiative in 2004, Dallas, Texas, partnered with Kirkuk, Iraq, later supplying seven thousand pairs of shoes and ten thousand pairs of socks to Iraqi orphans. San Diego, California, partnered with Jalalabad, Afghanistan, under the initiative and established educational sharing among elementary, middle, and high schools of both cities; the exchanges included both educational materials and the building of a new school in Jalalabad. Such exchange programs continue today, bringing together religious stakeholders for Sister City activities that reflect current community needs and interests. In 2015, a vibrant Sister City sewing circle at First African Methodist Episcopal Church in Seattle, Washington, decided to create washable feminine care kits to enable girls in Limbé, Cameroon, to attend schools during their menstrual cycles (SCI 2018).[3] The program partnered with Congregation Beth Shalom and Baitul Ehsaan Mosque, expanded its programming to address water-shortage issues in their community, and was providing additional water sanitation initiatives.

The Role of Cities in Civic Diplomacy

In the broader perspective of global-to-local ("glocal") activities, civic diplomacy corresponds to a suite of approaches for cities and municipalities in carrying out their foreign policy. As captured in Table 1, these include city diplomacy actions around specific place-based global engagement, municipal foreign policy, city twinning, city networks, and nonmunicipal city networks.

Ayres (2018) notes a growing new city multilateralism that is addressing key emerging issues such as climate change. She notes that city diplomacy is establishing itself as distinct from "paradiplomacy" and "constituent diplomacy," which connote privileged elite employed by national governments. City diplomacy is being implemented by citizen groups such as the United

---

3. See Seattle-Limbe Sister City Association, Seattle Limbe Sewing Circle, for further information (https://www.seattlelimbe.org)/.

Table 9.1. City Diplomacy Approaches and Characteristics

Type of City Diplomacy	Key Characteristics
Place-based global engagement	City diplomacy is not necessarily conducted in the halls and forums of international affairs; it happens in cities as international actors engage with local markets, politics, and cultures.
Municipal foreign policy	More cities take international engagements more seriously, outlining a more explicit "foreign policy" for municipal governments.
City twinning and bilateral relations	City-to-city bilateral relations are important modes of engagement by cities internationally, regionally, and nationally, and cities have a long history of collaboration.
City networks	City networks represent a mix of place-based engagements and municipal foreign policy, connecting local innovation with international collaboration.
Nonmunicipal city networking	International networking includes broader city diplomacy and globalizing connections between city hall and the world.

SOURCE: Adapted from Michele Acuto, Hugo Decramer, Juliana Kerr, Ian Klaus, Sam Tabory and Noah Toly, *Toward City Diplomacy: Assessing Capacity in Select Global Cities*, The Chicago Council on Global Affairs, July 7, 2018. Accessed July 10, 2018, https://www .thechicagocouncil.org/research/policy-brief/toward-city-diplomacy-assessing-capacity -select-global-cities.

Cities and Local Government organization, the Commonwealth Local Government Forum, the Global Parliament of Mayors, and the C40 Cities Climate Leadership Group.

One indication of this emerging role for cities as actors in international diplomacy is the emergence of protocol or international affairs offices in some cities. Cities such as New York, Paris, and San Antonio, Texas, have formed such offices with the express purpose of managing their foreign policy initiatives and transnational relations with other cities, nongovernmental and intergovernmental organizations, and nation-states (Rimmer and Ekanayake 2018). These offices may assist with civic diplomacy programs, be involved in transnational trade and economic initiatives, and organize trade missions and visits from delegations from other cities or countries. Some of these transnational cooperation opportunities may focus on key emerging issues such as smart energy, transnational crime, peace, health, security, and sustainability.

In this vein, city diplomacy encapsulates a number of key factors: at a municipal level, citizens are frontline actors—they may be organized in a decentralized way and may include a variety of implementers working together. Citizens as frontline actors represent the "troops on the ground," with a significant investment in "citizen or sidewalk diplomacy," serving an international diplomatic function for their community. The result is that regular city residents carry out international engagement as they partake in mutually beneficial international exchange programs.

This organization represents decentralized action at a local level. Leffel and Acuto (2017) highlight the "criticality of the networked topography of city diplomacy" (11) to emphasize the important functions that decentralized city networks play in dealing with complex issues such as climate change, particularly in an era of fractured national and international diplomacy efforts.

City multilaterals—voluntary institutions and associations, often working with local governments—carry out the activities. The emergence of cities and local entities as a driving force in global affairs has assumed increasing importance in many countries in what Curtis (2016) describes as the "transnational extension of national interests" (22). This role is vital. Tjandradew and Marcotullio (2009) note that cities and local governments have been developing international collaborative agreements for decades and indicate that more than 70 percent of the world's cities have international cooperative agreements with other cities. Increasingly, the critical importance of these relationships is re-emerging in light of the need for assistance and learning for common problems and challenges; global flows of information, technologies, people, and knowledge; and decentralization and new forms of intragovernment relations (see Server and Gruskin 2011).

Resilience: One of the Key Opportunities for City Diplomacy

Since the 1970s cities have been emerging as key actors in international affairs. This role is tied to a number of factors including urbanization, globalization, and decentralization (Nijman 2016). Others have noted a number of key trends in terms of internationalization of cities. Abbott (1997), for example, examined three distinct historic trends tied to these factors including the development of different kinds of cities. These include "production cities," which focus on manufacturing for world markets through exportation of finished goods and the establishment of branch plants; "gateway cities," which served as entry points for immigration, providing areas for cultural and economic international connectivity; and "transnational cities," which supply professional expertise, financial services, and personal services to multinational markets.

With such trends comes an appreciation of the possibilities that cities offer regarding "glocal" governance. Chan (2016) makes the case that in a twenty-first–century world of interdependence, state-to-state dialogues and diplomacy may be stalled by politics and gridlock, whereas city-to-city networks provide an opportunity for positive engagement. He notes that "cities share the common, pressing challenges in climate change, immigration, urban sustainability, housing, urban poverty, public health, security, jobs, and many more."

Chan (2016) identifies some benefits of the city-to-city approach in dealing with such issues. They include the proximity of cities to citizens, allowing them to serve as active participants in the ongoing political process; greater public trust at a municipal level than at a national level, because "close at hand" structures give citizens the experience of direct participation and influence; the ability of mayoral power to engage in global issues; and the possibility of cities to take specific action with or without formal mayoral power, particularly when it comes to climate action. As noted by Nijman (2016), former New York City mayor Michael Bloomberg, chair of the C40 Cities Climate Leadership Group, stated in 2012 that cities were well positioned to help address urgent problems because "we're the level of government closest to the majority of the world's people. We're directly responsible for their wellbeing and their futures. So, while nations talk, but too often drag their heels—cities act."

In the early 2000s, climate change resilience has emerged as one of the critical issues that cities and municipalities must address. Cities are now interacting with intergovernmental organizations in a new era of global action that has been developing since the Rio Earth Summit in 1992. Additionally, a focus on cities has emerged from the UN Sustainable Development Goals, as articulated in Goal 11, which is targeted at making cities inclusive, safe, resilience and sustainable (Rimmer and Ekanayake 2018).

This call to action at the city level for international engagement became even more salient when US president Donald Trump announced his intention to withdraw the United States from the Paris Agreement on climate change. As a result, the United States Climate Alliance was formed, bringing together state governments committed to upholding the Paris Agreement. Similarly, the United Nations special envoy for cities and climate change, Michael Bloomberg, submitted to the United Nations a statement of support entitled, "We Are Still In," from more than one thousand stakeholders, including business leaders, mayors, and governors (Engstrom and Weinstein 2018). Initiatives such as the Climate Alliance help build social movements that have been characterized as "transnational municipal networks," which are thought "to improve urban capacity to fight climate change and coordinate global urban efforts to that end" (Leffel and Acuto 2017, 10).

At its heart, resilience means leveraging the mechanisms and processes that strengthen the most vulnerable sectors of society and that can persist in times of crisis and disruption (De Souza 2018). Programs based on the concept of resilience aim to address environmental shocks—for example, from flooding, tornadoes, and earthquakes—and support rebuilding work following such shocks. They require short- and long-term inputs related to finance, planning, materials, resources, and a wide range of expertise and are likely to involve both local and national governments and whole communities (De Souza 2014).

A number of key resilience-building principles have emerged. One key principle focuses on adaptive capacities—that is, the ability of individuals in a community to deal with shocks, based on levels of exposure (the magnitude and frequency of shocks or degree of stress) and levels of sensitivity (the degree to which a system or actor is affected). The adaptive capacity of individuals, communities, regions, governments, organizations, and institutions is determined by the ability to adjust, to moderate the damage, to take advantage of opportunities, and to cope with the consequences. Another principle is the encouragement of a dynamic process of innovation and transformation. This approach to "learning as you go" is a cornerstone of resilience programming and allows for greater inclusion of innovation and receptivity to emerging needs.

There is increasing recognition of the importance of twinning efforts, which feed these resilience adaptive capacities and learning principles. "Learning cities," through urban neutral knowledge networks, benefit from traditional institutional knowledge and connections that arise in ad hoc and opportunistic ways. This premise, put forward and proved by Patrick Henry Buckley, Akio Takahashi, and Amy Anderson (2015), suggests that networking and knowledge sharing lead to improved quality of life and sustainability. Such improvements build a knowledge base and social capital to be shared across cities.

Others such as Stafford (2016) have commented on the value of the "soft power" of diffused political momentum that mayors use to act on climate change. Soft power conjures up an ability to achieve desired outcomes through legitimization versus coercion or payments.

Efraim Sitijak, Saut Sagala, and Elisabeth Rianawati (2014) make the case that cities could enhance their resilience dividend by engaging in Sister City relationships to catalyze the marketplace for resilient innovations and technologies; by developing hubs of learning and innovation; and by sharing social capital as encapsulated in twinning relationships with other cities that share similar threats. They specifically propose that Jakarta could focus on flood management and low carbon growth development through twinning relationships with Rotterdam in the Netherlands and Seoul in South Korea. The Sister City model of international engagement and diplomacy provides opportunities for addressing key international issues— such as climate resilience, that has local impact and where local diplomacy is emerging—as entry points for action.

The Role of Strategic Engagement in Glocal Movements

For nations to work together, to be informed, and to be rational actors on the international stage, individuals and organizations must take it upon themselves to experience and learn about one another. One key component of this approach is strategic engagement that implies getting to know and understand the perspectives of the key stakeholders in the local situation. Trent (2018) posits that an important contextual variable for city diplomacy is strategic engagement that emphasizes inclusive participation of concerned publics in problem-focused interactions. Sister City exchanges, and the Sister Cities model in general, are predicated on this concept of strategic engagement. Through the work of sister cities, peace is built through mutual respect, open communication, cooperation, and long-term engagement—one individual, one community at a time. Communications and engagement are part and parcel of this approach and are necessary because of the differences across communities and nations.

In traditional diplomacy, a country presents policies or viewpoints that are the formal position of the national government; in citizen diplomacy, there is recognition that all societies are composed of a mosaic of views across a spectrum and that engagement with all parts of society is needed to build understanding and trust. Sister City programs promote engagement because the differences between cultures and worldviews are real and often great (based on conversations in 2019 with Adam Kaplan, formerly of SCI).

As Trent (2018) notes, "Strategic diplomatic engagement using citizen and public diplomacy public private partnerships (P3s) that are rigorously planned, implemented and evaluated serves global-to-local (a.k.a. 'glocal') interests" (6). She concludes that partnerships demonstrate how participatory communication fosters cost-effective, inclusive networking and relationship building in public and citizen diplomacy. This occurs through intergenerational, cross-cultural, and socially responsible transnational-to-community-level mediating institutions such as SCI.

Key to the approach of strategic engagement in public diplomacy is the active, meaningful, and deliberate inclusion of civil society actors whose sphere of influence, experience, and values resonate with the goals of public diplomacy. This alignment is at the heart of the civic diplomacy approach. An important audience in this approach is diaspora communities, which are often active in Sister City relationships and leverage their relevant glocal perspectives and experience for effective engagement around a mutually conceived, specific policy problem or need. Similarly, religious actors—with their worldview and social capital—present opportunities for achieving mutual objectives.

Religion as Strategic Citizen Engagement

Historically in the United States, efforts to involve religious actors in public diplomacy have been housed in the State Department's (2021b) Office of International Religious Freedom. These efforts have been implemented through training programs offered by the US Agency for International Development (USAID) and the State Department's Office of Religion and Global Affairs. These programs were helping to engage with faith-based actors to advance secular developmental and diplomatic foreign policy goals (Trent 2016, 11). Such efforts typically seek to expand and foster interfaith understanding, pluralism, dialogue, and conflict mediation (Trent 2016, 13; Kovach 2016). Such efforts have experienced stops and starts (Casey 2017). As Kovach (2016) notes, "Religion is inexorably becoming a factor in our diplomacy and public diplomacy. But that marriage is very much a work in progress" (146).

Broadly defined, religion in public diplomacy may incorporate religious beliefs, worldviews, practices, and institutions across borders, time, and scale (Haluza-DeLay 2014). Efforts to include worldview and social organization have included bringing religious leaders from other countries to see interfaith collaboration in the United States or programs in countries like Bangladesh where USAID brought Hindu, Buddhist, Christian, and Muslim leaders together to promote action for the social good, such as positive health-seeking behaviors like visiting clinics (Kovach 2016, 165).

These public diplomacy, religious engagement efforts have also included work on climate resilience. As more interfaith groups have called for action on climate change, there is greater recognition of the need to mobilize congregations (Posas 2007). Outreach efforts from religious leaders have included the encyclical letter *Laudato Si' of the Holy Father Francis on Care of Our Common Home* for Catholics, the Church of England's General Synod, support statements from Hindu and Muslim leaders, a Buddhist climate change declaration, and a rabbinic letter on climate change (Pope Francis 2015; United Nations 2015). Stephen Goldsmith, William B. Eimicke, and Chris Pineda (2006) noted, however, that engaging faith-based organizations is not an automatic process. Some scholars suggest that faith-based organizations typically participate in programs that meet only short-term emergency needs such as food, clothing, and shelter. Others indicate that there may be negative community perceptions if services are provided to congregants over non-congregants, and some faith-based organizations may fear compromising their values if they work with local governments.

Moreover, some politically conservative evangelical Christian organizations in the United States have opposed and denied climate change (Haluza-DeLay 2014). Others identify several barriers for engaging religious actors in climate actions. These may be summarized in four types of barriers: paradigmatic (beliefs that disable environmental engagement), applicable (doubt about the appropriate amount of attention to give environmental concerns), critical (inadequate attention to social or cultural factors as they affect faith or environmental matters), and convictional (lifestyle and willingness to act) (Haluza-DeLay 2008). Some Christians, for example, may believe that God will protect them—this may lessen their efforts to take prevention and adaptive measures against climate change (Ives 2016; Pozniak 2016).

Complementarity between Religious Activism and Citizen Diplomacy

Despite such barriers, many believe that there are benefits to aligning religious activism with citizen diplomatic efforts. Haluza-DeLay (2014) indicates that religions may influence believers' worldviews or cosmologies and the moral duties that they promote. In addition, religions may engage a broad audience, many members of which accept and respect religious moral authority and leadership. He also notes that religions could leverage significant institutional and economic resources at their disposal and provide social capital to leverage positive climate action. Bakker (2011) makes the case that faith-based, partnership-oriented development efforts, like sister church relationships, carry much promise—for example, they are generally trusted by their communities, they create and provide community leadership, they have access to human and financial capital in the form of volunteers and donations, they are community and cultural anchors in areas where they have long been located, and they are typically more readily holistic in nature.

Berry (2018) has identified various ways that the religious community may play a role in the public sphere and on emerging issues such as the environment. He notes that religion "can advance or resist processes of democratization, support or neglect issues of social justice, reinforce or erode state sovereignty. In other words, religion can act sometimes as a counter-hegemonic force and sometimes as a legitimizing force" (38). In this context, he examines ways that transnational religious organizations have served as key actors in environmental and climate advocacy efforts in Latin America and proposes that social justice is an important organizing platform. He affirms that that "climate change—or more accurately, the proliferating institutional responses to climate change—serve as a nexus for organizations pursuant of social justice" (44).

With these potential benefits in mind, there are a few common bases of religious strategic engagement in climate change action. The first is a commonality of the human experience: "Christians, Muslims, Hindus, Buddhists and Jews may have different beliefs, but they all share the same resources and they are all affected by the consequences of climate change" (Deutsche Welle 2021). As part of this shared experience, there may be a recognition that practical way to love one's neighbor is to help each other through shared experiences.

Second, some suggest that in particular geographies (e.g., small island developing states), no progress will be made on a purely secular climate change agenda. Patrick D. Nunn (2017), from the University of the Sunshine Coast in Australia, argues that secular projects in the islands are failing: "Unless you are cocooned in a tourist bubble, it is hardly possible to miss God when you visit the Pacific Islands. In every village and on every main street there seems to be a church or temple, packed to bursting point on holy days. . . . Yet almost very well-intentioned outside agency—including those of foreign governments . . . that seeks to help the region's people adapt to the effects of future climate change is drawing up its plans in secular ways, and [plans are] communicated using secular language."

Third, calls for action by religious leaders are premised on a civic duty to act. Such calls to action focus on areas around justice and equity, stewardship, and service to others. Some ways that this is articulated to religious actors is to recognize that such actors are called "to be part of God's restorative work in the world, restoring key relationships; between God and humankind; between humankind itself (across race, gender, language, ability, etc.) and humankind with the rest of creation" (Hyneman and Shore 2013; see *Nassau Guardian* 2015).

Finally, religious norms are believed to be much more in line with how citizens experience and navigate their everyday lives than are the secular norms that shape climate policy discussions. Religious practices connect individuals to the reality of climate changes in their lives by allowing them to experience these changes first hand. This is often articulated in terms of a stewardship ethic: "Humans are unique in our call to be stewards of creation. The way we live and treat creation impacts all people, including ourselves, and we are called to care for it, as a part of our expression of God's image in us" (Hyneman and Shore 2013; see *Nassau Guardian* 2015).

International Intercongregational Relationships

Interestingly, the last quarter of twentieth century saw a rise of faith-community relationships between developed and developing nations, called "international intercongregational relationships" (Cosgrove, Fogg, and Moore 2010). These relationships have interchangeable names—twinning, sister, or partnership—and different levels of sister-community involvement. These levels may include, at a minimum, charities; at an intermediate level, community development programs; or at a systemic level, projects that address far-reaching issues such as climate change. International intercongregational relationships have included eco-parishes, eco-twinning, eco-congregations, and sister parishes.

Gable and Haasl (2016) suggest that Catholic twinning processes and parishes have evolved with greater congregation involvement because of easier and faster communications. These processes stress solidarity over charity, wherein the twinning mission model should promote mutuality (both parties benefit from the relationship) and solidarity, movement away from colonial attitudes, and a greater emphasis on long-term sustained missions (see Gable and Haasl 2016; Vos, Kithikii, and Pagnucco 2008).

There are many instances of sister parish efforts to build resilience through mitigating and adapting to climate change. The Church of England, for example, encourages tree planting overseas to combat climate change (*Telegraph* 2009). Similarly, Our Sister Parish (2021) of Des Moines focuses on adaption methods to climate change for agriculture in El Salvador, whereas the Global Catholic Climate Movement (2021) makes parallel efforts. The movement serves the Catholic family worldwide, encouraging it to transform Pope Francis's Laudato Si' into action for climate justice by undergoing ecological conversion, transforming lifestyles, and calling for bold public policies together with the wider climate movement. The Global Catholic Climate Movement represents a coalition of more than four hundred global Catholic member organizations that include Caritas agencies, religious orders, lay movements, youth groups, diocesan offices, and Catholic-inspired nongovernmental organizations as members.

Eco-twinning is an approach led by the United Nations and the Alliance of Religions and Conservation (ARC, n.d.). "Twinning" is the practice of towns, parishes, schools, or dioceses developing long-lasting relationships of mutual benefit between communities in two distinct places. "Eco-twinning" takes that idea further, linking churches in the global North with those in the global South that are experiencing the detrimental effects of climate change first hand. Parishes in the global North provide funds to a matched church in the global South. Virtual and web-based exchanges build the relationship. Lessons in church schools are tailored to environmental lessons learned from their sister parish; those with fewer experiences of the direct immediate impacts of climate change have the opportunity to see them through their sister parish (ARC 2021). Similarly, there are examples of eco-congregations that have emerged to share similar faith-based obligatory notions to protect nature (Jones 2016).

Opportunities for Mutual Benefits and Impact

To develop robust and resilient responses to climate change, climate diplomacy needs to be "strategically inclusive" and build partnerships among all the various stakeholders. Religious actors are an important stakeholder group whose inclusion in civic and climate diplomacy promises specific value. Moving forward, there are potential areas for collaboration and engagement that take these conclusions even further. These opportunities include promoting action from a shared ethic of global stewardship, maximizing emerging governance structures for shared goals around climate resilience, presenting and engaging in actions around a holistic approach to resilience, and leveraging the shared assets of key resilience actors in the religious and diplomatic communities.

One of the key areas for further collaboration of citizen diplomacy, religion, and climate resilience is around an ethic of global stewardship. Many Christians are being called to action, for example, around a restoration of relationships to engage in God's restorative work in the world, acknowledging key relationships between God and humankind and between humankind and creation. This call corresponds to being called to be stewards of creation and of fellow men, getting to know God through serving others (United States Conference of Catholic Bishops 1997). One possible approach might be to change the ways in which climate change is framed. "At its core, global climate change is . . . about the future of God's creation and the one human family. It is about protecting both 'the human environment' and the natural environment. It is about our human stewardship of God's creation and our responsibility to those who come after us," the United States Conference of Catholic Bishops said in its 2001 statement (see Catholic Climate Covenant 2018). Some of the primary goals of Catholic activism on climate change aim at changing perceptions of climate change through a lens of four Christian principles: acting with prudence in the face of uncertainty, protecting the poor and most vulnerable, working toward a common good, and protecting human solidarity and future generations (United States Conference of Catholic Bishops 2001).

The ethics of stewardship, mutual interests and understanding, and shared experiences are at the heart of the approach of civic diplomacy. In the context of climate resilience, such ethics recognize the vulnerabilities of affected communities and promote collaboration with them to identify strengths and capacities to increase shared resilience, especially at times of disaster.

Furthermore, exploring emerging religious and civic governance structures could help accelerate climate-smart innovations. Kalim Shah has identified new governance mechanisms for climate resilience in the Caribbean that have enabled communities to build resilience by leveraging different types of capital.[4] Social capital has been recognized as being particularly important for developing resilience. Shah identifies an active and vibrant parish council system in Barbados that leverages the local governance structures through traditional parish church and village assembles. This type of new governance structure for resilience is one that engages religious communities and provides benefits for cross border learning on resilience models and approaches.

Spickard (2020) also noted the important role of important role of organizers and pastoral agents who used such structures together with church-based organizing to mobilize a congressional base by identifying leaders among the congregations and training and empowering them.

There may also be ways to use diplomacy to build a holistic approach to resilience. Speaking in 2018 at a forum organized by SCI and American University, Kate Bentley explained that cultural diplomacy has created a space for collaboration and negotiation across borders.[5] Bentley served as a US negotiator on environmental issues in an era of mistrust and strong national positions. She described that when she was able to connect with her foreign counterparts in informal spaces through song, they all found commonality in shared cultural experiences, which led to an opening up of zones of negotiation. She noted, "The foundations built through cultural diplomacy are, indeed, invaluable as it bridges divides and transcends political barriers. And most importantly, they promote mutual understanding and respect among countries and its citizens."

Recognizing the importance of cultural diplomacy in civic movements alludes to some of the roots of resilience thinking, harkening back to the personal resilience of individuals during crises. Religious communities recognize and engage people's emotional and spiritual needs as being key to their resilience. Religious institutions and leaders have a role in providing pastoral, prayerful contributions to personal and communal well-being.

---

4. Lynae Bresser references Kalim Shah, assistant professor of environmental policy at Indiana University, in reporting on the "Developing Climate Resilience: An Island Perspective" event, part of the Managing Our Planet Series from the Wilson Center, October 5, 2016 (www.wilsoncenter. org/node/49466).

5. Presented at "Diplomacy, Religion, and Resilience in Small Island Developing States," May 22, 2018, Washington, DC (https://www.american.edu/centers/latin-american-latino-studies/upload /SCI_Climate-Brochure_R2.pdf).

As communities deal with disasters and recovery, there is an opportunity for alignment with community and religious contributions to human well-being.

Finally, resilience actors could leverage shared assets that advance mutual goals. Religious institutions and citizen diplomats bring assets to work that builds resilience. These assets include religious places of worship such as church buildings, as well as community centers, schools, social networks, methods of mass communication, and volunteers for risk reduction and preparedness. There are also soft assets such as trusted relationships, respect and authority in the community, pastoral and emotional support, and prayer and spiritual guidance. Both communities also can represent the case of the most vulnerable to responsible authorities.

Conclusion

This chapter outlined ways in which civic diplomacy generally, and citizen diplomacy specifically, is an increasingly important dimension of climate diplomacy. This role is tied to the dimensions of the glocal structure of environmental challenges and to factors that drive action at a local level, such as national governments' abdication of leadership on climate. The Sister City model of citizen diplomacy contributes to city diplomacy with citizen diplomats for their communities. To develop robust and resilient responses to climate change, climate diplomacy needs to be strategically inclusive and build partnerships among all stakeholders. This diplomacy is linked to key concepts and approaches such as strategic engagement, mutuality, and solidarity. Including religious actors in citizen diplomacy and climate diplomacy promises specific areas for collaboration and engagement that take these conclusions even further.

The tools of diplomacy and religion involve creating a space for meaningful engagement. Shared values of mutual respect, reciprocity, and service are important opportunities for mutual interests. The time is right for these movements to come together. In a time of fractured national politics, twinning movements, religious actors, and citizen diplomats share similarities and opportunities for engagement. There has been some exploration of nascent approaches through international intercongregational relationships, but there is greater scope through mutual framing of the issues, providing services and innovations that cross borders while leveraging common assets and promoting a shared opportunity for global stewardship.

References

Abbott, Carl. 1997. "The International City Hypothesis: An Approach to the Recent History of US Cities." PhD thesis, Portland State University. Accessed November 1, 2018. https://core.ac.uk/download/pdf/37771149.pdf.

ARC (Alliance of Religions and Conservation). n.d. "Eco-Twinning." http://www.arcworld .org/downloads/ARC-Eco-Twinning-leaflet.pdf.

———. 2021. "Projects: Eco-Twinning." http://arcworld.org/projects.asp?projectID=367.

Ayres, Alyssa. 2018. "The New City Multilateralism." Council on Foreign Relations, June 27, 2018. Accessed September 1, 2018. https://www.cfr.org/expert-brief/new-city -multilateralism.

Bakker, Janel Kragt. 2011. "Molding Mission: Collective Action Frames and Sister Church Participation." *Review of Faith and International Affairs* 9 (3): 11–20. doi:10.1080/155702 74.2011.597204.

Berry, Evan. 2018. "Transnational Religious Advocacy Networks in Latin America and Beyond." In *Church, Cosmovision, and the Environment: Religion and Social Conflict in Contemporary Latin America*, edited by Even Berry and Robert Albro, 37–54. Routledge Studies in Religion and Environment. London: Routledge.

Buckley, Patrick Henry, Akio Takahashi, and Amy Anderson. 2015. "The Role of Sister Cities' Staff Exchanges in Developing 'Learning Cities': Exploring Necessary and Sufficient Conditions in Social Capital Development Utilizing Proportional Odds Modeling." *International Journal of Environmental Research and Public Health* 12 (7): 7133–53.

Casey, Shaun. 2017. "How the State Department Has Sidelined Religion's Role in Diplomacy." *Religion and Politics*, September 5, 2017. Accessed June 30, 2021. https:// religionandpolitics.org/2017/09/05/how-the-state-department-has-sidelined-religions -role-in-diplomacy/.

Catholic Climate Covenant. 2018. "Catholic Climate Declaration." Accessed September 2, 2021. https://catholicclimatecovenant.org/files/inline-files/CatholicClimateDeclaration Leave-BehindDocument_0.pdf.

Chan, Dan Koo-hong. 2016. "City Diplomacy and 'Glocal' Governance: Revitalizing Cosmopolitan Democracy." *Innovation: The European Journal of Social Science Research* 29 (2). https://doi.org/10.1080/13511610.2016.1157684.

Chung, Mona, and Bruno Mascitelli. 2009. "A New Dimension of Sister City Relationships in the 21st Century: A Pilot Study in Australia." Global Business and Technology Association.

Cosgrove, John, Doug Fogg, and Ellen Moore. 2010. "The Sister Parish Phenomenon." Presented at the North American Association of Christians in Social Work Convention, November 2010, Raleigh-Durham, NC. Accessed August 22, 2018. https:// www.nacsw.org/Publications/Proceedings2010/CosgroveJSisterParish.pdf.

Cox, Aubrey, Melissa Nozell, and Imrana Alhaji Buba. 2017. *Implementing UNSCR 2250: Youth and Religious Actors Engaging for Peace*. Washington, DC: United States Institute of Peace. Accessed April 30, 2020. https://www.usip.org/sites/default /files/2017-06/sr406-implementing-unscr-2250-youth-and-religious-actors-engaging -for-peace.pdf.

Curtis, Simon. 2016. "Cities and Global Governance: State Failure or a New Global Order?" *Millennium* 44 (3): 455–477.

De Souza, Roger-Mark. 2014. "Resilience, Integrated Development and Family Planning: Building Long Term Solutions." *Reproductive Health Matters* 22 (43): 75–83.

———. 2018 "The More Things Change: Resilience, Complexity, and Diplomacy Are Still Top Priorities in 2018." *New Security Beat* (blog), January 2, 2018. Accessed August 2, 2018. https://www.newsecuritybeat.org/2018/01/change-resilience-complexity-diplomacy -top-priorities-2018/.

Deutsche Welle. 2021. "Environment: Faith and Climate Protection." https://www.dw.com /en/top-stories/green-technology/s-100334.

Engstrom, David Freeman, and Jeremy M. Weinstein. 2018. "What If California Had a Foreign Policy? The New Frontier of States' Rights." *Washington Quarterly* 41 (1): 27–43. https://doi.org/10.1080/0163660X.2018.1445356.

Feffer, John, and Nancy Snow. 2016. "Anti-Americanism and the Rise of Civic Diplomacy." *Foreign Policy in Focus*, December 13, 2016. https://fpif.org/anti-americanism_and _the_rise_of_civic_diplomacy/.

Gable, Mike, and Mike Haasl. 2016. "Training for the Third Wave of Mission: A Catholic Perspective." Working Papers of the American Society of Missiology, Vol. 2. Accessed August 22, 2018. https://place.asburyseminary.edu/cgi/viewcontent.cgi?article=1064&c ontext=firstfruitspapers.

Global Catholic Climate Movement. 2021. "Catholics Protect Creation." Accessed June 30, 2021. https://catholicclimatemovement.global/introduction/.

Goldsmith, Stephen, William B. Eimicke, and Chris Pineda. 2006. *Faith-Based Organizations versus Their Secular Counterparts: A Primer for Local Officials.* Cambridge, MA: Ash Institute for Democratic Governance and Innovation John F. Kennedy School of Government, Harvard University. Accessed August 22, 2018. https://www.innovations .harvard.edu/sites/default/files/11120.pdf.

Grote, Dora. 2012. "Chinese Vice President Xi Jinping Visits Muscatine." *The Daily Iowan*, February 17, 2012. Accessed on June 30, 2021. https://international.uiowa.edu/news /chinese-vice-president-xi-jinping-visits-muscatine.

Haluza-DeLay, Randolph. 2008. "Churches Engaging the Environment: An Autoethnography of Obstacles and Opportunities." *Human Ecology Review* 15 (1): 71–81.

———. 2014. "Religion and Climate Change: Varieties in Viewpoints and Practices." *WIREs Climate Change* 5 (2): 261–279. https://doi.org/10.1002/wcc.268.

Hyneman, Jared, and Christopher Shore. 2013. *Why Are We Stewards of Creation? World Vision's Biblical Understanding of How We Relate to Creation.* Uxbridge, UK: World Vision International. Accessed August 22, 2018. https://www.wvi.org/sites/default/files /World%20Vision's%20Biblical%20Understanding%20of%20How%20we%20Relate%20 to%20Creation_Full.pdf.

Ives, Mike. 2016. "A Remote Pacific Nation, Threatened by Rising Seas." *New York Times*, July 2, 2016. https://www.nytimes.com/2016/07/03/world/asia/climate-change-kiribati .html.

Johnson, Kirk. 2012. "For the Vice President of China, Tea Time in Iowa." *New York Times*, February 15, 2012. Accessed June 30, 2021. https://www.nytimes.com/2012/02/16/world /asia/xi-jinping-of-china-makes-a-return-trip-to-iowa.html.

Jones, Rick. 2016. "From North Carolina to Scotland: Earth Care Congregation Partnership Reaps Benefits." Presbyterian News Service, December 28, 2016. https://www .presbyterianmission.org/story/north-carolina-scotland-earth-care-congregation -partnership-reaps-benefits/.

Kovach, Peter. 2016. "Public Diplomacy Engages Religious Communities, Actors, and Organizations: A Belated and Transformative Marriage." In *Nontraditional US Public Diplomacy: Past, Present, and Future*, edited by Deborah L. Trent, 145–70. Washington, DC: Public Diplomacy Council.

Leffel, Benjamin, and Michele Acuto. 2017. "City Diplomacy in the Age of Brexit and Trump." *City Diplomacy, Public Diplomacy* 18 (Summer/Fall): 13.

Maselko, Joanna, Cayce Hughes, and Rose Cheney. 2011. "Religious Social Capital: Its Measurement and Utility in the Study of the Social Determinants of Health." *Soc Sci Med* 2011 73 (5): 759–767. doi:10.1016/j.socscimed.2011.06.019.

Nassau Guardian. 2015. "Antilles Episcopal Conference Declaration on Climate Change." BahamasLocal.com, June 10, 2015. https://www.bahamaslocal.com/newsitem/128047 /Antilles_Episcopal_Conference_declaration_on_climate_change.html.

Nijman, Janne Elisabeth. 2016. "Renaissance of the City as Global Actor: The Role of Foreign Policy and International Law Practices in the Construction of Cities as Global Actors." T.M.C. Asser Institute for International and European Law 2016-02. Updated June 30, 2017. https://papers.ssrn.com/sol3/papers.cfm?abstract _id=2737805.

Nunn, Patrick. 2017. "Sidelining God: Why Secular Climate Projects in the Pacific Islands Are Failing." *Conversation*, May 16, 2017. Accessed August 27, 2017. http:// theconversation.com/sidelining-god-why-secular-climate-projects-in-the-pacific -islands-are-failing-77623.

Our Sister Parish. 2021. "What We Do." Accessed June 30, 2021. http://oursisterparish.org /what-we-do.htm.

Pew Research Center. 2019. "Religion's Relationship to Happiness, Civic Engagement and Health around the World." Accessed June 30, 2021. https://www.pewforum.org/2019 /01/31/religions-relationship-to-happiness-civic-engagement-and-health-around-the -world/.

Pope Francis. 2015. "Laudato Si'of the Holy Father Francis on Care of Our Common Home." Vatican. Accessed June 30, 2021. http://w2.vatican.va/content/francesco/en/encyclicals /documents/papa-francesco_20150524_enciclica-laudato-si.html.

Posas, Paula J. 2007. "Roles of Religion and Ethics in Addressing Climate Change." *Ethics in Science and Environmental Politics* 2007:31–49. https://www.int-res.com/articles /esep/2007/E80.pdf.

Pozniak, Kim. 2016. "Praying for Rain in Ethiopia." Catholic Relief Services, July 1, 2016. https://www.crs.org/stories/praying-rain-ethiopia.

Rimmer, Susan Harris, and Charuka Ekanayake. 2018. "New Rules for the Neighbourhood? The Rise and Rise of Asia's Cities." In *State of the Neighbourhood 2018*, edited by Caitlin Byrne and Lucy West, 35–47. Southport, Australia: Griffith University. Accessed November 10, 2018. https://www.griffith.edu.au/__data/assets/pdf_file/0024/583404 /State-of-the-Neighbourhood-2018.pdf.

SCI (Sister Cities International). 2006. *Peace through People: 50 Years of Global Citizenship*. Louisville, KY: Butler.

———. 2018. "Sister Cities Stitch Together Washington and Cameroon." Accessed June 30, 2021. https://sistercities.org/2018/05/24/sister-cities-stitch-together-washington-and-cameroon/.

———. 2021. "Our History." Accessed June 30, 2021. http://sistercities.org/about-us/our-mission/our-history/.

Server, Daniel, and Peter Gruskin. 2011. "Municipal Twinning for the 21st Century." Internal report to Sister Cities International on its Global Sister Cities and Twinning Summit, Cairo, September 7–9, 2011. Johns Hopkins School for Advanced International Studies, October 6, 2011.

Sitinjak, Efraim, Saut Sagala, and Elisabeth Rianawati. 2014. "Opportunity for Sister City Application to Support Resilience City." Working Paper Series, No. 8. Resilience Development Initiative. Accessed August 15, 2018. https://www.preventionweb.net/files/39758_39758wp8sitinjaketalopportunityfors.pdf.

Spickard, James V. 2020. "Social Capital, Civic Capital: Local Churches Organize for Popular Democracy." Working Papers and Reports, University of Redlands. Accessed June 30, 2021. https://inspire.redlands.edu/work/ns/906562f1-a30a-4e10-bafd-58cfdc2ddc96.

Stafford, Max. 2016. "Ironic Power Shifts: The Emergence of Mayors as Users of Soft Power within Global Governance." Presented at the Political Studies Association annual conference, University of Sussex, Brighton, March 23, 2016. Accessed September 1, 2017. https://www.psa.ac.uk/sites/default/files/conference/papers/2016/PSA%20Paper%20 2016%20Pdf.pdf.

Telegraph. 2009. "Church of England to Encourage Tree Planting to Combat Climate Change." *Telegraph*, November 3, 2009. https://www.telegraph.co.uk/news/earth/environment/climatechange/6487029/Church-of-England-to-encourage-tree-planting-to-combat-climate-change.html.

Tjandradewi, Bernadia Irawati, and Peter J. Marcotullio. 2009. "City-to-City Networks: Asian Perspectives on Key Elements and Areas for Success." *Habitat International* 33:165–72.

Trent, Deborah L., ed. 2016. *Nontraditional US Public Diplomacy: Past, Present, and Future.* Washington, DC: Public Diplomacy Council.

———. 2018. *Many Voices, Many Hands: Widening Participatory Dialogue to Improve Diplomacy's Impact.* Los Angeles: Figueroa.

United Nations. 2015. "Islamic Declaration on Climate Change." United Nations Climate Change, August 18, 2015. Accessed September 2, 2021. https://unfccc.int/news/islamic-declaration-on-climate-change.

United States Conference of Catholic Bishops. 1997. "Called to Global Solidarity International Challenges for US Parishes." Statement, November 12, 1997. Accessed September 2, 2021. http://www.usccb.org/issues-and-action/human-life-and-dignity/global-issues/called-to-global-solidarity-international-challenges-for-u-s-parishes.cfm.

———. 2001. "Global Climate Change: A Plea for Dialogue Prudence and the Common Good." Statement, June 15, 2001. Accessed September 2, 2021. http://www.usccb.org/issues-and-action/human-life-and-dignity/environment/global-climate-change-a-plea-for-dialogue-prudence-and-the-common-good.cfm.

US Department of State. 2021a. "Bureau of Educational and Cultural Affairs." https://www.state.gov/bureaus-offices/under-secretary-for-public-diplomacy-and-public-affairs/bureau-of-educational-and-cultural-affairs/.

———. 2021b. "Office of International Religious Freedom." https://www.state.gov/bureaus
-offices/under-secretary-for-civilian-security-democracy-and-human-rights/office-of
-international-religious-freedom/.

Vos, William, Agnes Kithikii, and Ron Pagnucco. 2008. "A Case Study in Global Solidarity:
The St. Cloud-Homa Bay Partnership." *Journal for Peace and Justice Studies* 17 (1):
45–60. Accessed September 2, 2021. https://www.csbsju.edu/Documents/Peace%20
Studies/pdf/Global-Solidarity-Updated.pdf.

ROGER-MARK DE SOUZA is Chief Movement Building Officer at
Amnesty International USA. Previously he served as president and
CEO of Sister Cities International (SCI). His most recent publications
on resilience include "Building Resilience for Peace: Water, Security,
and strategic Interests in Mindanao, Philippines," in *Water, Security,
and US Foreign Policy* (2017); "Beautiful, Vulnerable, Innovative:
Challenges and Opportunities for the Caribbean as Incubators of
Responsible Innovation," in *Coastal Tourism, Sustainability and
Climate Change in the Caribbean* (2017); and "Demographic Resilience:
Linking Population Dynamics, the Environment and Security,"
The SAIS Review of International Affairs 35, no.1 (2015) 17–27.

CONCLUSION

Where Climate Meets Religion—Mobilization, Discourse, and Authority

Ken Conca

To a scholar who has done extensive research on social movements, mobilization, and political activism, often with an empirical focus on water politics and policy, the role of religion is at once both intimately familiar and quite alien. On one hand, religion intersects fundamentally with several of the core concepts that help us understand how movements arise, do what they do, and have an impact on the world. Consider the concept of resource mobilization, used to examine movements in terms of "the variety of resources that must be mobilized, the linkages of social movements to other groups, the dependence of movements on external support for success, and the tactics used by authorities to control or incorporate movements" (McCarthy and Zald 1977, 1213). Religion provides a set of social networks through which connections are made, resources are mobilized, and political activism is conducted. It is no accident that religious activism was central to movements as distinct as the civil rights campaigns in the United States in the 1960s, peasant and indigenous groups organizing across Latin America in the 1980s, and engagements in Myanmar during the recent political transition. In each instance, institutionalized social networks—the Southern Christian Leadership Conference, the leftist "liberation theology" wing of the Catholic Church, or the Buddhist *sangha* monastic orders (Frydenlund 2017; Kurzman 1998)—solved what can be debilitating problems for movements: reducing the costs of organizing, facilitating the flow of information and ideas, and enabling strategic coordination.

Similarly, consider framing, another central concept in social movement studies. *Framing* has been defined as "the signifying work or meaning construction engaged in by movement adherents (e.g., leaders, activists, and rank-and-file participants) and other actors (e.g., adversaries, institutional elites, media, social control agents, counter-movements) relevant to the interests of movements and the challenges they mount in pursuit of those interests" (Snow 2013, 470). Religious themes and imagery provide potent tools for activists seeking to express collective identities. They may also be useful in the essential task of "frame alignment"—the process of presenting a movement's aims and values as consistent with existing normative frameworks in the wider community. Environmental protection becomes an expression of God's expectation of stewardship over His domain on Earth. Hizballah campaigns for political change in Lebanon become guidance for "devout Muslims who desire to follow God's path . . . a prognostic frame offering a divine solution to Lebanon's problems" (Karagiannis 2009, 375). Campaigns for a shorter work week become part of an obligation to protect the sabbath (Mirola 1999). Although none of these alignments is easy, uncontested, or without second-order consequences, the potential synergies are obvious. Indeed, John Hannigan (1991) has argued that processes of contestation, globalization, and empowerment provide a common foundation for social movement theory and the sociology of religion.

Nevertheless, there is also a substantial disconnect between religion and the conceptual tools commonly used to study movements. Part of the problem is theoretical: Browers's (2005, 75) charge of a secular bias in distinctions between ideology ("defined as political, action-orientated and intimately tied to modernity") and religion ("dismissed as otherworldly, conservative and anti-modern") could be extrapolated to any of the key concepts in social movement theory. An equally vexing problem is empirical: case studies of environmental activism, for example, tilt heavily toward groups with progressive, modernist, secular, and globalist worldviews. Selection bias is also a challenge at the level of data collection—for example, reliance on newspaper accounts to assemble "events" data (Hug and Wisler 1998) or working in places that are most easily accessible, linguistically and culturally, to the researcher (Hendrix 2017). These filters put us at risk of missing a broad swath of activity that occurs in the name of faith and of force-fitting what they do allow us to see into culturally particularistic categories.

Interestingly, a similarly jarring dichotomy is found in another area where I have substantial research experience—the study of water politics and policy. Across the world's major religions, water is a recurring and multidimensional symbol, tied to the origins of life, purification, spiritual cleansing, renewal, and passage to the next life or a new state of being (Oestigaard 2009). Chamberlain (2008) has argued that it is essential to listen to the stories emanating from these traditions if we are to construct a "new water ethic," distinct from the dominant ethos of development, centralization, marketization, and technocratic managerialism. He argues that insights from religion are an essential resource to make "a sharp break with the tenacious hold of a mechanistic, utilitarian, materialistic view of the natural world" (6). While not informed by an such quest, development bureaucracies such as the US Agency for International Development (USAID) and the World Bank have also sought to instrumentalize the religion–water interface, bringing religious actors into the conversation around water aid and development programming (see, e.g., USAID 2013).

Consider this statement by an activist with the Highlands Church Action Group of Lesotho explaining her opposition to dam construction by summarizing the role of the river in Besotho life:

> A river plays a very big role in our culture. It has a lot to do. If somebody passes away or maybe was killed by the lightning, usually he would be buried next to the river. It is a place where our traditional doctors go to get qualified. Some people say they talk with their ancestors right in the river. If a girl is about to start her first period, a traditional way to guide her is to take her to the river. Apart from that, if someone in the family dreams about a river, it will mean that someone in the family is pregnant; and if I am a mother, I should know that something is wrong with one of my daughters. (IRN 1997, 77)

This way of knowing about water and rivers, and this way of being with them in the world, is jarringly different from standard accounts of antidam activism driven by environmental impact assessments and human rights criticism. Similarly, most policy studies on extending water and sanitation services to those who lack them have failed to note the substantially different practices used by secular and religious aid providers (and between the more secularly oriented and the more religiously oriented groups within the religious aid community) (McLeigh 2011; Goldsmith, Eimicke, and Pineda 2006). Studies critical of marginal cost pricing and privatization may object on grounds of human rights, social justice, or technical efficacy but rarely note the Koran's teaching that water should be freely available and that surplus water is not to be withheld. Consequently, when studying the intersection of religion and climate change, we face the challenge of using the theoretical and empirical tools at our disposal while remaining acutely aware of their limits and shortcomings.

The following core questions animate the chapters in this volume: First, how are religious actors bringing their concerns around climate change to bear? Second, how do religious ideas or frameworks shape how actors make sense of climate change as a socio-ecological phenomenon? Third, how is climate change itself driving religious change? In the sections that follow, I will draw out several comparative observations about how the chapters speak to these questions, organized around themes of mobilization, discourse, and authority. In doing so, I draw insights from the social science literature on these themes while trying to remain sensitive to the historical construction of that literature, which has rendered it a less-than-perfect toolkit for examining religion and religious actors. My strategy is to tack between theory and observation—to take observations from the chapter and ask, "What is this an instance of?" while remaining open to possibilities that the patterns found in the chapters' ethnographic accounts may not conform to the expectations of a largely secular world of eco-political theorizing.

When we see religion providing an organizing vehicle for actors entering the political arena, we must ask whether it serves as just another form of institutional infrastructure for mobilizing political action—or a particularly efficacious one—or something more. Similarly, when we see actors using religious teachings and practices as interpretive frames to make sense of climate change, does this constitute one way of knowing among many? Or a particularly significant way of knowing? Or something more? The goal is not to narrowly segment the chapters' observations into existing social science variables but rather to engage in ways of thinking that can generate more refined questions for a climate–religion research agenda. There is a danger that such an approach forcibly fits the chapter observations into pre-existing categories (if the chapter authors themselves have not already done so). Still, this approach can be useful for climate–religion scholarship to be in conversation with both the social science of climate change and contemporary policy dialogues about adaptation and mitigation. The fine-grained character of the chapters' case accounts lends itself to this approach.

How Are Religious Actors Entering the Spaces of Climate Politics?

Religious actors face several obstacles in entering the political arena. Wilson (chap. 4) notes that secular ontologies prevail in some of the most important political spaces for climate engagements. Those spaces include the polities of the leading emitters of greenhouse gases, the international machinery of climate governance, and the networked activities of mainstream nongovernmental organizations. In addition, the urban and modernist biases in all these arenas constitute significant obstacles for many of the actors captured in this volume. As Berry (chap. 5) puts it, "The invitation to religion is an invitation to partnership in political projects already underway." Moreover, as De Souza (chap. 9) notes, many faith-based organizations that consider entering the political arena remain wary of congregant perceptions, paradigmatic tensions, and the compromises entailed in working with secular authorities (see also Goldsmith et al. 2006).

The idea of previously nonpolitical actors organizing to engage secular political processes and institutions belies an important observation found in several of the chapters—that the space for political engagement on climate change is already structured by both religious beliefs and the institutionalization of religious–political channels of influence. In some cases, the "political projects already underway" bear a strong, prior religious stamp. In the Philippines case (chap. 1), mainstream religious actors enjoy longstanding, privileged access to the corridors of power. In India (chap. 6), the political tensions around river pollution and state water governance programs are shaped by the increasingly muscular projections of Hinduism that have washed over that country's national political institutions. In Trinidad (chap. 2), the long-standing, embedded notion that "God is a Trini" supports political influence for those preaching the gospel of prosperity (and the status quo of resource extractivism). In each of these cases, what Buckley (chap. 1) terms "the institutional relationship between religion and the state" emerges as an important piece of the opportunity structure. In other cases, the significance of this variable is seen in the negative; local communities, finding themselves shut out of national conversations about resource extraction in Peru (chap. 7) or disaster risk in Puerto Rico (chap. 8), mobilize in different directions and prioritize local engagements.

The idea of opportunity structure, however, may mislead if it suggests fixed or immutable characteristics of these political systems. The political agency of religious actors is alive and well in these cases, and how they choose to engage with (dis)advantageous opportunity structures is a significant part of each story. In Berry's paired-opposite examples of conservative American Christians and the Vatican of Pope Francis, climate change is instrumentalized in both cases to follow the paths of least (political) resistance and maximum influence. In the Philippines, the environmental concerns of the religious community are subordinated to the larger tensions of engaging a problematic regime.

Not surprisingly, in several cases, faith-related social channels provide well-institutionalized networks of solidarity, trust, and shared values through which collective agency can flow more easily. Nevertheless, as these actors mobilize through such channels, a striking part of their stories is the degree to which "sampling" or selective tapping of religious beliefs and traditions occurs. Catholic priests in Peru borrow from indigenous cosmology in a way that strengthens their role as agents of protection from *wa'ka* vengeance. Vedwan (chap. 3) discusses how a range of actors in India, from Hindu nationalists to middle-class urban dwellers to environmental activists, draw selectively on religious–cultural traditions to make sense of pollution, climate change, and environmental responsibility. Elite agents also engage in the selective sampling of religious traditions: both Pope Francis and India's Modi regime draw selectively from historical cultural–religious themes to frame their international positions on climate change. Dryzek's (2005) caution that "discourses have to be treated as less totalizing and constraining than some followers of Michel Foucault claim" (224) also applies to faith-derived discourses and points us to the importance of framing.

Do Religious Actors and Their Frames Change the Discursive Politics of Climate Change?

As Berry (chap. 5) notes, "As the moral and spiritual dimensions of the climate crises become increasingly evident, and as the limitations of scientific information as a source of social transformation become painfully clear, powerful voices have championed the contributions and potential of religion." Such interest in religion is driven not only by the capacity of religion to (de)mobilize actors in the political arenas of climate mitigation or adaptation but also by how religion may change the conversation therein. De Souza (chap. 9), for example, sees potential contributions to the construction of a "shared ethic of global responsibility" and a "holistic approach to resilience." Similarly, Vedwan (chap. 3) notes the "underutilized" role of religion in creating "a discourse that is scientifically robust, culturally salient, economically equitable, and politically inclusive."

The qualification that religion be understood not simply as a set of beliefs or principles, antecedent to social action around them, is paramount. As De Souza notes in chapter 9, religious political engagements "may incorporate religious beliefs, worldviews, practices, and institutions across borders, time, and scale." It is particularly important to keep this multidimensionality in mind when studying discursive politics and framing processes: practices may speak as loudly as ideas, and institutional or cultural commitments may trump theological ones.

A second important qualification is that framing is not simply the application of ideas to a cause but rather the deliberate construction of an interpretive lens, or "the conscious strategic efforts by groups of people to fashion shared understandings of the world and of themselves that legitimate and motivate collective action" (McCarthy 1997, 244). Snow et al. (1986) identified three distinct functions performed by acts of framing that they labeled *diagnostic, prognostic,* and *motivational.* Diagnostic frames identify problems and assign blame; prognostic frames suggest responses; and motivational frames provide a rationale for action.

We see some innovative framings across the cases that could speak to current forms of contention or blockage in climate politics. One is the idea, developed most extensively in Crosson's chapter on Trinidad (chap. 2) but seen in other cases as well, of the injured Earth as environmental–theological construct that sits, so to speak, between Gaia and stewardship. As Crosson notes, the Gaian construct demotes human agency, severing responsibility in a way that can be profoundly demobilizing, whereas stewardship offers an uninspiring "middle management" form of human responsibility. Animating the landscape through the notion that "the earth is the Lord" activates the idea of an injured earth and a role for human empathy. Crosson reads this as an innovative balance between the excessive separation of human agency within stewardship and the problematic amalgamation of that agency in a Gaian framework.

Broadly speaking, the idea of an injured earth triggering an empathetic (or at least similarly balanced) response can also be seen in several other cases. In the Peruvian case, there is a clear parallel in Bacigalupo's observation that "neither a radical ontological approach nor a skeptical modernist one is useful for understanding how poor mestizos in La Libertad engage with wak'as for political ends." In Puerto Rico, accepting inherent vulnerability but finding in it the social and cultural resources for a more sustainable and resilient way of living is also directly analogous. In India, middle-class residents of Delhi occupy potentially similar ground when they make sense of environmental degradation through narratives that stress the loss of traditions of personal responsibility and communal harmony. In each case, the moral relationship between people and planet is more complex and dynamic than either reducing them to the "forcing function" of climate change within the planetary system, on one hand, or extracting them as the freestanding custodian of subordinate nature, on the other.

A second innovative frame involves the idea of vulnerability. Within the international climate governance regime, engagements around the question of vulnerability have proven to be some of the most politically nettlesome. One of the principal ways in which recognition of vulnerability is expressed in global political discourse is through the contentious debate around climate change as a security issue, which has been batted around in the United Nations (UN) Security Council for a decade now with little to show for it (Conca 2019). Within the UN Framework Convention on Climate Change, serious engagements with the question of vulnerability have also run aground—in this case, due to the politics of liability around "loss and damage." One may debate the efficacy of securitizing or financializing climate change as a rhetorical move or an agenda-setting tactic. Ultimately, both discourses prove to be profoundly conservative in terms of how they conceptualize that which is ultimately vulnerable and, therein, the values at risk. One seeks to insulate existing sovereign–political relations and the stability of nation-states from the perturbing effects of climate impacts, and the other seeks to preserve the economics of business-as-usual through the premise of compensation for climate-driven losses. It is not that these framings lack, in Crosson's terms, "agency, power, and claims for justice," but rather that the terms in which they constitute agency, exercise power, and seek justice work primarily to protect the status quo and blunt our recognition of the full scope of the problem.

Thompson's discussion of vulnerability in the context of Puerto Rico offers a very different construct. In this case, vulnerability is in part inherent and in part rooted in the failings of those very institutions; and it is understood to highlight opportunity and capacity as well as risk and loss. While religion is not the only source for a more transformative approach to vulnerability—the secular side of the humanitarian-aid regime has quite clearly retooled itself in this direction over the past few decades—it provides actors in several of the cases with such a path. Poor mestizos in Peru use their vulnerability (to climate change and the structural violence of extractivism) to fashion creative new alliances and a collective ethic of environmental responsibility. Working-class Trinis, in a very different context, are engaged in a broadly similar project. Various actors in India, including ordinary citizens, connect their understanding of vulnerability to resources available to them in Indian culture.

There are also important limits to the power of such frames, however innovative or transformative their potential. Most obvious is the problematic politics of scale around climate change. Physical scale imposes some clear limits: the changing global climate seems poised to punish Peru, no matter the efficacy of appeasing the wak'a. Equally problematic are the effects of scaling up these grounded sociocultural frames for making sense of climate change in the context of the modern administrative state. Both case studies on India demonstrate such scale effects. In chapter 6 by Alley and Mehta on river politics, grounded practices of understanding a sacred river as a living entity are contradicted (or at least riven with tensions) by the larger political–administrative context with which this idea is translated into legal personhood. Vedwan's discussion in chapter 3 of how everyday Indians use cultural traditions to piece together an understanding of environmental change shows stylized "tradition" playing very differently when in the hands of the Modi regime on a global stage. What serves a transformational agenda at one scale may have different effects at another.

What Is the Basis for Authority of Religious Interventions in Climate Politics?

In all but the most nakedly power-based conceptions of politics, authority is a key concept. Authority is the marriage of power to legitimacy—actors enjoy authority when the application of power to seek a given end is met by other parties with legitimacy in the form of some minimal degree of acknowledgement, acceptance, and willing compliance (Lake 2003; Scheppele and Soltan 1987). The ability to exercise authority, in turn, is the key to political order.

The legitimization of power as authority may occur through one or more of several mechanisms. Many scholars have found it useful to distinguish between "input" and "output" legitimacy, with the former referring to the process by which decisions are reached and actions are defined and the latter to some results-oriented validation of those decisions or actions. In the former, my authority as a classroom instructor is rooted in my doctorate, the careful vetting when I was hired, and the training I have received; in the latter, it is the demonstrated capacity to teach well. In practice, authority is typically a blend of the two. Consider science-based policy advocacy, which may enjoy some degree of authority rooted in the technical expertise of scientists. Such authority is based on both the ability to validate scientific claims empirically (output legitimacy) and the reputation of science and the scientific method (input legitimacy), derived from a history of accrued validations.

Importantly, there is no single basis for the authority of political actors, particularly for those who sit outside the formal institutions of the state. Authority may be grounded in technical expertise ("I can build a dam that will generate electricity and won't collapse"), a compelling moral claim ("I speak for God's will"), or a plausible claim of representation ("I speak for future generations"). Authority may also derive from delegation, as when UN member-states infuse a UN organ or specialized agency with the legitimized power to take actions on public health, development assistance, or peacekeeping interventions.

Interestingly, several case studies suggest a basis for the authority of religious beliefs, actions, or practices other than the obvious candidate of moral authority. In Peru, grassroots religious actors blend tales of morality with multiple bids for output legitimacy, rooted in both formalized science and the "science" of their own grounded experience. As Bacigalupo notes in chapter 7, this hybridized basis for authority of their claims about climate allows them to connect to a broader audience: "*curanderos . . .* and visionaries . . . predict floods, mudslides, and tsunamis and using different discourses of authority to legitimate their predictions and to make them accessible to a large interethnic audience." In Trinidad, something similar occurs as working-class Trinis blend a theological interpretation of an injured earth with their own empirical observations about flooding, storm exposure, and climate change (although, unlike the hybridized narratives in Peru, Crosson notes that such sense-making remains unintelligible to Western climate science). In both instances, the amorality of the animated landscape—the selfish wa'kas and the vengeful, injured earth—focuses attention on cause-and-effect understandings.

In both cases, the role of belief systems is more to help people make sense of change—a process related to output legitimacy—than simply to codify moral law. The deliberative element of such sense-making suggests another potential pathway to legitimation. As the authority of the nation-state has grown fragmented by processes of social, economic, and technological change, it is no accident that there has been a resurgence of interest in "stakeholder dialogue" and deliberative approaches to policy making, particularly around highly contentious issues (Mena and Palazzo 2012). Thompson (chap. 8) posits that the concept of vulnerability can play a bridging role in such deliberative dialogue. Crosson (chap. 2) goes further: "Might the notion that the earth can become unsettled and respond to human actions be a point at which both natural science and theology are partially converging?"

Other chapters caution against the notion of pluralist ontological dialogue, much less (partial) convergence. Wilson's discussion of secular ontologies reminds us that such convergence retains problematic elements, both ontologically and institutionally. Vedwan notes that, at least in India, one such bid for convergence has been highly contentious, as science and technology have become key battlegrounds on which "ideologues seek to assert and advance a cultural and religious revivalist agenda." Also in India, the faith-driven status of legal personhood for rivers, while serving specific agendas of powerful actors, holds for case authors Alley and Mehta little promise of performance-based legitimacy in cleaning up rivers, reducing them instead to "Hinduized icons of political and administrative agendas."

How Does Climate Change Affect Religion?

As Jenkins, Berry, and Kreider (2018) have noted, "Understanding of religion is changing as its relations with climate comes into view." Historians have begun to link epochal shifts in climate to changes in religious thought and practice. If climate change proceeds at the pace most climate models suggest, then such work is not finished. Jenkins and colleagues, however, also note the importance of taking a reflexive approach to the relationship: "Even if not always in predictable, linear patterns, and not always in explicit response to climate change, the basic idea here is that anthropogenic changes in human-natural systems exert reflexive pressure on the religious systems through which, in turn, many persons interpret the meaning of global environmental change. That reflective interpretive spiral may drive cosmological shifts analogous in scope to those of the Axial Age" (98). Moreover, such reflexivity must apply not simply to the observations of climatic change but also to the political struggles and moral controversies surrounding responses to it, from renewable energy programs to pipeline activism to geoengineering.

Given the focus of the chapters on emergent political agency about (and amid) climate change, it is beyond the scope of this brief synthesis to make more than tentative observations about religion as the dependent variable. In addition, Vedwan (chap. 3) reminds us that climate policies and practices are not seamless and fully realized projects but rather assemblages—fraught with inconsistency, ambiguity, heterogeneity, and reversal—rendering any predictions about their impacts difficult.

Nonetheless, one pattern that stands out across the chapters follows the observation, noted earlier, by Berry (chap. 5): "The invitation to religion is an invitation to partnership in political projects already underway." As faith-guided actors mobilize, construct interpretive frames, and seek legitimacy for the power they wield, they also become associated with those political projects. The state and the market founder in their attempts to formulate projects that simultaneously address climate change and reauthorize the state and the market—and they do so in a world where 80 percent of people practice a religion, and half of all schools are religiously affiliated. It should be little surprise, then, that people seek answers in religious thought and practice. In an era of profound political polarization, where the answers themselves have been commoditized—where knowledge is a good sought to validate comfortable beliefs rather than a foundation on which to seek truths—there is also significant peril in that quest.

References

Browers, Michaelle L. 2005. "The Secular Bias in Ideology Studies and the Case of Islamism." *Journal of Political Ideologies* 10 (1): 75–93.

Chamberlain, Gary. 2008. *Troubled Waters: Religion, Ethics, and the Global Water Crisis.* Lanham, MD: Rowman & Littlefield.

Conca, Ken. 2019. "Is There a Role for the UN Security Council on Climate Change?" *Environment* 61 (1): 4–15.

Dryzek, John. 2005. "Deliberative Democracy in Divided Societies: Alternatives to Agonism and Analgesia." *Political Theory* 33 (2): 218–42.

Frydenlund, Iselin. 2017. "Religious Liberty for Whom? The Buddhist Politics of Religious Freedom during Myanmar's Transition to Democracy." *Nordic Journal of Human Rights* 35 (1): 55–73.

Goldsmith, Stephen, William B. Eimicke, and Chris Pineda. 2006. "Faith-Based Organizations versus Their Secular Counterparts: A Primer for Local Officials." Cambridge, MA: Ash Institute for Democratic Governance and Innovation, John F. Kennedy School of Government, Harvard University. Accessed June 29, 2021. https://www.innovations.harvard.edu/sites/default/files/11120.pdf.

Hannigan, John A. 1991. "Social Movement Theory and the Sociology of Religion: Toward a New Synthesis." *Sociology of Religion* 52 (4): 311–31.

Hendrix, Cullen. 2017. "The Streetlight Effect in Climate Change Research on Africa." *Global Environmental Change* 43:137–47.

Hug, Simon, and Dominique Wisler. 1998. "Correcting for Selection Bias in Social Movement Research." *Mobilization* 3 (2): 141–61.

IRN (International Rivers Network). 1997. *Proceedings, First International Meeting of People Affected by Dams, Curitiba, Brazil, March 11–14, 1997.* Berkeley, CA: International Rivers Network.

Jenkins, Willis, Evan Berry, and Luke Beck Kreider. 2018. "Religion and Climate Change." *Annual Review of Environment and Resources* 43:9.1–9.24.

Karagiannis, Emmanuel. 2009. "Hizballah as a Social Movement Organization: A Framing Approach." *Mediterranean Politics* 14 (3): 365–83.

Kurzman, Charles. 1998. "Organizational Opportunity and Social Movement Mobilization: A Comparative Analysis of Four Religious Movements." *Mobilization* 3 (1): 23–49.

Lake, David A. 2003. "The New Sovereignty in International Relations." *International Studies Review* 5:303–23.

McCarthy, John D. 1997. "The Globalization of Social Movement Theory." In *Transnational Social Movements and Global Politics: Solidarity beyond the State*, edited by in Jackie Smith, Charles Chatfield, and Ron Pagnucco, 243–59. Syracuse, NY: Syracuse University Press.

McCarthy, John D., and Mayer N. Zald. 1977. "Resource Mobilization and Social Movements: A Partial Theory." *American Journal of Sociology* 82 (6): 1212–41.

McLeigh, Jill. 2011. "Does Faith Matter? A Comparison of Faith-Based and Secular International Nongovernmental Organizations Engaged in Humanitarian Assistance." PhD diss., Clemson University.

Mena, S., and G. Palazzo. 2012. "Input and Output Legitimacy of Multi-stakeholder Initiatives." *Business Ethics Quarterly* 22 (3): 527–56.

Mirola, William. 1999. "Shorter Hours and the Protestant Sabbath: Religious Framing and Movement Alliances in Late Nineteenth-Century Chicago." *Social Science History* 23 (3): 395–433.

Oestigaard, Terje, ed. 2009. *Water, Culture and Identity: Comparing Past and Present Traditions in the Nile Basin Region.* Bergen, Norway: BRIC.

Scheppele, Kim Lane, and Karol Edward Soltan. 1987. "The Authority of Alternatives." In *Authority Revisited: NOMOS XXIX*, edited by J. Roland Pennock and John W. Chapman, 169–200. New York: New York University Press.

Snow, David A. 2013. "Framing and Social Movements." In *The Wiley-Blackwell Encyclopedia of Social and Political Movements*, edited by David A. Snow, Donatella Della Porta, Bert Klandermans, and Doug McAdam, 470–75. Hoboken, NJ: Wiley Blackwell.

Snow, David A., R. Burke Rochford, Jr., Steven K. Worden, and Robert D. Benford. 1986. "Frame Alignment Processes, Micromobilization, and Movement Participation." *American Sociological Review* 51:464–81.

USAID (US Agency for International Development). 2013. "Faith, Water and Development," *Global Waters* 4 (3): 1–20.

KEN CONCA is Professor of International Relations at American University. His research focuses on environment, conflict, and peacebuilding; water politics and governance; and the politics of climate adaptation. He is a two-time recipient of the International Studies

Association's Harold and Margaret Sprout Award for best book on international environmental affairs and a recipient of the Chadwick Alger Prize for best book in the field of international organization. In 2017, he was awarded the Al-Moumin Environmental Peacebuilding Lecture, which recognizes thought leaders on environmental change, conflict, and peace; in 2021, he received the Grawemeyer Award for Ideas Improving World Order. His most recent book is *The Oxford Handbook of Water Politics and Policy*. Conca earned his PhD from the Energy and Resources Group at the University of California, Berkeley.

INDEX

Page numbers in *italics* indicate figures.

Abbott, Carl, 238
Achuar people, 114
ACT Alliance, 21
Acuto, Michele, 238
Adger, William, 69
Adi Yogi, statue of, 158
Affordable Care Act (United States), 33
African Caribbean Spiritual Baptist
  religion, 41
African religions, 32, 45, 50, 57
Agard, John, 48
Agarwal, Anil, 81
Agarwal, G. D., 79–80
agency, 217, 262; moral, of sentient
  landscapes, 177–82, 185, 192–201
ahimsa (nonviolence), 87–88
Allen, George V., 233
Alley, Kelly D., 7, 8, 10, 87, 150, 263, 265
Alliance of Religions and Conservation
  (ARC), 246
Allison, Elizabeth, 218
Allocco, Amy, 67n1
Amazon, 180, 186, 196
American Psychological Association (APA),
  211, 218
Andean Incas (1470–1532), 184
Andean region, 5, 8, 177, 183–86, 197.
  See also La Libertad (Andean region,
  Peru)
Anderson, Amy, 240
animist religions, 44, 57
Annenberg Public Policy Center study, 137
Anthropocene, 1, 10, 60, 200; *kaliyuga*,
  compared with, 67, 67n1
anthropology, 7, 113, 181, 184
anthropomorphism, 60
antidevelopment scholarship, 187
antimodernist critiques, 69

anxieties, 1, 3, 68, 79, 89, 216
apocalyptic beliefs, 42, 52–54, 133, 197–200
apus (divine beings), 8, 176–201; *apu* as
  honorary title, 184; Cerro Campana, 177,
  188, 193–94, 196; Cuculicote, 179.
  See also sentient landscapes
Aquino, Benigno, 19, 24
archaeologists, and sentient landscapes,
  191–92, 198
Arguelles, Ramon (Archbishop), 25–26
Art of Living Foundation (AOL), 153–58,
  167, 170
Asad, Talal, 106, 111
Asociación de Rescate y Defensa del Apu
  Campana (Association for the Rescue and
  Defense of Apu Campana), 193
assemblages, climate discourses as, 60,
  67, 69, 266
Association of Major Religious Superiors of
  the Philippines, 18
atmospheric system, borderlessness of, 5
austerity, 46–47
austerity gospel, 42, 50–51; "prosperity-
  austerity gospel," 42–43, 51–52
Australia, 97, 123–24; Indigenous peoples,
  113–14, 116
authority, 249, 258, 263–65; moral, 177, 194, 244
Ayres, Alyssa, 236
Ayurveda, 75

Bacigalupo, Ana Mariella, 7, 8, 264
Baitul Ehsaan Mosque, 236
Bakker, Janel Kragt, 244
Banerjee, Sumanta, 85
Bangladesh, 243
Barbados, 248
Barnes, Jessica, 88
Bataan Nuclear Power Plant (Philippines), 26

nonsecular, 109, 112, 113, 115, 116; secular, 9, 99–118, 265
opportunity structures, 130–32, 140
"other," 10, 60, 89, 235
other-than-human beings, 7–8, 59. *See also* nonhuman actors; personhood
Oxford Companion to Christian Thought, 44–45, 46

Pachamama (Mother Earth), 179, 188, 195
Pacific Island communities, 127, 220
Palmer, Brian, 135–36
Paris Agreement, 124; COP 21, 124, 137, 138; and Holy See, 137; India's ratification of, 82; and Philippines, 15, 21, 25, 26; Trump's rejection of, 133–34, 239
Peace of Westphalia (1648), 110
Peace Park (Nagasaki, Japan), 232
Pentecostal-charismatic Christianity, 45–46
People-to-People Conference (1956), 230
per-capita emission allowance, 81–82
Persad-Bissessar, Kamala, 46, 263
personhood: of Amazon rain forest, 196; of animals, 114; of corporations, 196; of Earth, 59; of glaciers, lakes, and wetlands, 160; of nature, 194–96; persons in loco parentis, 159–60, 163, 169; of rivers, 8, 158–66, 169, 263; of sentient landscapes, 180–81. *See also* other-than-human beings; sentient landscapes
Peru, 7, 10, 176–205, 259; health problems caused by extractivism, 186, 191; serrano-coastal mestizo dichotomy, 182–83, 185; sugarcane industry, 183, 185, 186; vulnerability to climate change, 185–86. *See also* La Libertad (Andean region, Peru); mestizos, poor (Peru); sentient landscapes
petroleum: disasters caused by tropical storms and hurricanes, 48; fall of oil prices, 46–47; oil as blood of the Earth, 43, 52, 54; as "resource curse," 43
petroleum extraction: and apocalyptic predictions in Trinidad, 52–54; fishing and farming endangered by, 56; as injury

to God, 8, 43, 56–57, 61; as parasitic, 54, 56; Trinidad, 6, 42–43, 46; Trinidad as earliest site of, 56, 59
Pew Research Center study, 133
Philippine Council of Evangelical Churches (PCEC), 18, 31
Philippine Misereor Partnership Inc., 24
Philippines, 6, 9, 15–38, 222, 259–60; Bangsamoro Autonomous Region, 18; "benevolent secularism," 22–24, 32; civil society, 20–21, 24; coal-fired power plants, 26; Commission on Human Rights, 29; Constitution, 22–23, 29; Constitutional Commission, 18, 23; death penalty restoration proposal, 29, 30; democracy, 17–19, 31–32; demographics of religion, 17–18; Department of Energy and Natural Resources (DENR), 24, 25–26; elections, 15, 18, 24, 27–30; extrajudicial killing (EJK), 28–31; implications for transnational climate politics, 32–34; Indigenous communities, 20; institutions, importance of, 22–24, 32; logging ban, 23–24; Marcos regime, 18, 27; media, restrictions on, 30–31; Mindanao, 18, 20, 25, 27; mining interests, 20, 25, 27, 31; and Paris Agreement, 15, 21, 25, 26, 31; People Power Revolution (1986), 15, 17–18, 23, 29, 32; popular protests (2001), 18–19; religious, political, and environmental landscape of, 16–22; Reproductive Health Act of 2012, 19; Supreme Court, 23, 30; Typhoon Haiyan (Yolanda), 21–22; vulnerability to climate change, 19–20; "war on drugs," 16, 27–33
phreatic water layer, 198–99
physical-flows view/biophysical vulnerability model, 208, 210
Pineda, Chris, 243
planetary boundaries framework, 170–71
policy infrastructures, 8
political ideologies, 42n3, 62, 67, 129
political theology, 41
politics: as local, 6; myth of as rational, 102. *See also* cosmopolitics, transactional; global politics; nation-state; state

EVAN BERRY is Assistant Professor of Environmental Humanities in the School of History, Philosophy, and Religious Studies at Arizona State University. He is author of *Devoted to Nature: The Religious Roots of American Environmentalism.*

CPSIA information can be obtained
at www.ICGtesting.com
Printed in the USA
LVHW101631010622
720234LV00003B/70